PRAISE FOR *MYSQL CRASH COURSE*

"A fantastic resource for anyone who wants to learn about MySQL . . . and an excellent refresher for more seasoned developers."

—Scott Stroz, MySQL Developer Advocate

"Understand not just the 'what,' but the 'why' behind MySQL development."

—Steven Sian, web and mobile application developer

MYSQL CRASH COURSE

A Hands-on Introduction to Database Development

by Rick Silva

no starch press

San Francisco

Printed in the United States of America

First printing

27 26 25 24 23   1 2 3 4 5

ISBN-13: 978-1-7185-0300-7 (print)
ISBN-13: 978-1-7185-0301-4 (ebook)

Publisher: William Pollock
Managing Editor: Jill Franklin
Production Manager: Sabrina Plomitallo-González
Production Editor: Jennifer Kepler
Developmental Editors: Rachel Monaghan, Eva Morrow, and Frances Saux
Cover Illustrator: Gina Redman
Interior Design: Octopod Studios
Technical Reviewer: Frédéric Descamps
Copyeditor: Rachel Monaghan
Compositor: Jeff Lytle, Happenstance Type-O-Rama
Proofreader: Scout Festa

For information on distribution, bulk sales, corporate sales, or translations, please contact No Starch Press, Inc. directly at info@nostarch.com or:

No Starch Press, Inc.
245 8th Street, San Francisco, CA 94103
phone: 1.415.863.9900
www.nostarch.com

Library of Congress Cataloging-in-Publication Data

```
Names: Silva, Rick, author.
Title: MySQL crash course : a hands-on introduction to database development / Rick Silva.
Description: San Francisco, CA : No Starch Press, Inc., [2023] | Includes index.
Identifiers: LCCN 2022050277 (print) | LCCN 2022050278 (ebook) | ISBN 9781718503007 (print) | ISBN
   9781718503014 (ebook)
Subjects: LCSH: SQL (Computer program language) | MySQL (Electronic resource) | Computer
   programming.
Classification: LCC QA76.73.S67 S557 2023 (print) | LCC QA76.73.S67 (ebook) | DDC 005.75/6--dc23/
   eng/20221128
LC record available at https://lccn.loc.gov/2022050277
LC ebook record available at https://lccn.loc.gov/2022050278
```


[S]

To my wife, Patti, for her patience, love, and support. You are Mother Teresa in a scarf.

About the Author

Rick Silva is a software developer with decades of database experience. Silva has worked at Harvard Business School, Zipcar, and various financial services companies. A Boston native and a Boston College alum, he now lives in the Raleigh, North Carolina, area with his wife, Patti, and his dog, Dixie. When he's not joining database tables, he's playing banjo at a local bluegrass jam.

About the Technical Reviewer

Frédéric Descamps (@lefred) has been consulting on open source and MySQL for more than 20 years. After graduating with a degree in management information technology, he started his career as a developer for an ERP system under HP-UX. He then opted for a career in the world of open source by joining one of the first Belgian startups dedicated 100 percent to free projects around GNU/Linux. In 2011 Frédéric joined Percona, one of the leading MySQL-based specialists. He joined the MySQL Community Team in 2016 as a MySQL Community Manager for EMEA and APAC. Descamps is a regular speaker at open source conferences and a technical reviewer for several books. His blog, mostly dedicated to MySQL, is at *https://lefred.be.*

Descamps is also the devoted father of three adorable daughters: Wilhelmine, Héloïse, and Barbara.

BRIEF CONTENTS

CONTENTS IN DETAIL

4
MYSQL DATA TYPES 37

5
JOINING DATABASE TABLES 51

6
PERFORMING COMPLEX JOINS WITH MULTIPLE TABLES 63

11
CREATING FUNCTIONS AND PROCEDURES 165

12
CREATING TRIGGERS 195

13
CREATING EVENTS 209

PART IV: ADVANCED TOPICS 217

14
TIPS AND TRICKS 219

17
TRACKING CHANGES TO VOTER DATA WITH TRIGGERS 277

18
PROTECTING SALARY DATA WITH VIEWS 301

AFTERWORD 313

INDEX 315

ACKNOWLEDGMENTS

This book wouldn't have been possible without the team of professionals at No Starch Press, to whom I am deeply thankful. Thanks to Bill Pollock for believing in the concept and setting the ship on the right course.

I am grateful for the exceptionally talented team of editors at No Starch Press, including Rachel Monaghan, Eva Morrow, Frances Saux, Jenn Kepler, and Jill Franklin. Thanks to Miles Bond for his hard work on the book, and to Eric Matthes for taking the time to share his insights with me.

I want to thank Frédéric Descamps, affectionately known as @lefred in MySQL circles, for being the technical reviewer of this book. His attention to detail and deep knowledge of MySQL are impressive and appreciated.

Thanks to Jimmy Allen and the gang at Oasis for all the encouragement.

INTRODUCTION

In the mid-1980s, I landed my first software development job, which introduced me to the *relational database management system (RDBMS)*, a system to store and retrieve data from a database. The concept has been around since 1970, when E.F. Codd published his famous paper introducing the relational model. The term *relational* refers to the fact that the data is stored in a grid of rows and columns, otherwise known as a table.

At the time I started out, commercial database systems weren't widely available. In fact, I didn't know anybody else who was using one. The RDBMS I used was imperfect, with no graphical interface and a command line interface that periodically crashed for no apparent reason. Since the World Wide Web had yet to be invented, there were no websites I could turn to for help, so I had no choice but to start my system back up and hope for the best.

Still, the idea was pretty cool. I saved large amounts of data in tables I created based on the nature of the information I wanted to store. I defined table columns, loaded data into the tables from files, and queried that data with *Structured Query Language (SQL)*, a language for interacting with databases that allowed me to add, change, and delete multiple rows of data in a snap. I could manage an entire company's data using this technology!

Today, relational database management systems are ubiquitous and, thankfully, far more stable and advanced than the clunkers I used in the '80s. SQL has also vastly improved. The focus of this book is MySQL, which has become the most popular open source RDBMS in the world since its creation in 1995.

About This Book

This book will teach you to use MySQL using its Community Server (also known as the Community Edition), which is free to use and has the features most people need. There are also paid versions of MySQL, including the Enterprise Edition, that come with extra features and capabilities. All editions run on a wide variety of operating systems, such as Linux, Windows, macOS, and even the cloud, and have a robust set of features and tools.

Throughout this book, you'll explore the most useful parts of MySQL development, as well as insights I've picked up over the years. We'll cover how to write SQL statements; create tables, functions, triggers, and views; and ensure the integrity of your data. In the last three chapters, you'll see how to use MySQL in the real world through hands-on projects.

This book is organized in five parts:

Part I: Getting Started

Chapter 1: Installing MySQL and Tools Shows you how to download MySQL and offers some tips for installing it on various operating systems. You'll also install two tools to access MySQL: MySQL Workbench and the MySQL command line client.

Chapter 2: Creating Databases and Tables Defines databases and tables and shows how to create them. You'll also add constraints to your tables to enforce rules about the data they will allow and see how indexes can speed up data retrieval.

Part II: Selecting Data from a MySQL Database

Chapter 3: Introduction to SQL Covers how to query database tables to select the information you want to display. You'll order your results, add comments to your SQL code, and deal with null values.

Chapter 4: MySQL Data Types Discusses the data types you can use to define the columns in your tables. You'll see how to define columns to hold strings, integers, dates, and more.

Chapter 5: Joining Database Tables Summarizes the different ways you can select from two tables at once, covering the main types of joins and how to create aliases for your columns and tables.

Chapter 6: Performing Complex Joins with Multiple Tables Shows you how to join many tables as well as use temporary tables, Common Table Expressions, derived tables, and subqueries.

Chapter 7: Comparing Values Walks you through comparing values in SQL. For example, you'll see ways to check whether one value is equal to another, greater than another, or within a range of values.

Chapter 8: Calling Built-in MySQL Functions Explains what a function is, how to call functions, and what the most useful functions are. You'll learn about functions that deal with math, dates, and strings, and use aggregate functions for groups of values.

Chapter 9: Inserting, Updating, and Deleting Data Describes how to add, change, and remove data in your tables.

Part III: Database Objects

Chapter 10: Creating Views Explores database views, or virtual tables based on a query you create.

Chapter 11: Creating Functions and Procedures Shows you how to write reusable stored routines.

Chapter 12: Creating Triggers Explains how to write database triggers that automatically execute when a change is made to data.

Chapter 13: Creating Events Shows you how to set up functionality to run based on a defined schedule.

Part IV: Advanced Topics

Chapter 14: Tips and Tricks Discusses how to avoid some common problems, support existing systems, and load data from a file into a table.

Chapter 15: Calling MySQL from Programming Languages Explores calling MySQL from within PHP, Python, and Java programs.

Part V: Projects

Chapter 16: Building a Weather Database Shows you how to build a system to load weather data into a trucking company's database using technologies such as cron and Bash.

Chapter 17: Tracking Changes to Voter Data with Triggers Guides you through the process of building an election database, using database triggers to prevent data errors, and tracking user changes to data.

Chapter 18: Protecting Salary Data with Views Shows you how to use views to expose or hide sensitive data from particular users.

Every chapter includes "Try It Yourself" exercises to help you master the concepts explained in the text.

Who Is This Book For?

This book is suitable for anyone interested in MySQL, including folks new to MySQL and databases, developers who would like a refresher, and even seasoned software developers transitioning to MySQL from another database system.

Since this book focuses on MySQL *development* rather than *administration*, MySQL database administrators (DBAs) may want to look elsewhere. While I occasionally wander into a topic of interest to a DBA (like granting permissions on tables), I don't delve into server setup, storage capacity, backup, recovery, or most other DBA-related issues.

I've designed this book for MySQL beginners, but if you'd like to attempt the exercises in your own MySQL environment, Chapter 1 will guide you through downloading and installing MySQL.

SQL in MySQL vs. SQL in Other Database Systems

Learning SQL is an important part of using MySQL. SQL allows you to store, modify, and delete data from your databases, as well as create and remove tables, query your data, and much more.

Relational database management systems other than MySQL, including Oracle, Microsoft SQL Server, and PostgreSQL, also use SQL. In theory, the SQL used in these systems is standardized according to the American National Standards Institute (ANSI) specifications. In practice, however, there are some differences among the database systems.

Each database system comes with its own extension of SQL. For example, Oracle provides a procedural extension of SQL called Procedural Language/SQL (PL/SQL). Microsoft SQL Server comes with Transact-SQL (T-SQL). PostgreSQL comes with Procedural Language/PostgreSQL (PL/pgSQL). MySQL doesn't have a fancy name for its extension; it's simply called the MySQL stored program language. These SQL extensions all use different syntaxes.

Database systems created these extensions because SQL is a *non-procedural* language, meaning it's great for retrieving and storing data to or from a database, but it isn't designed to be a procedural programming language like Java or Python that allows us to use `if...then` logic or `while` loops, for example. The database procedural extensions add that functionality.

Therefore, while much of the SQL knowledge you learn from this book will be transferable to other database systems, some of the syntax may require tweaking if you want to run your queries with a database system other than MySQL.

Using the Online Resources

This book includes many example scripts, which you can find at *https://github.com/ricksilva/mysql_cc*. The scripts for Chapters 2–18 follow the naming convention *chapter_X.sql*, where *X* is the chapter number. Chapters 15 and 16 have additional scripts in folders named *chapter_15* and *chapter_16*.

Each script creates the MySQL databases and tables shown in the corresponding chapter. The script also contains example code and answers for the exercises. I recommend attempting the exercises yourself, but feel free to use this resource if you get stuck or want to check your answers.

You can browse through the scripts and copy commands as you see fit. From GitHub, paste the commands into your environment using a tool like MySQL Workbench or the MySQL command line client (these tools are discussed in Chapter 1). Alternatively, you can download the scripts to your computer. To do this, navigate to the GitHub repository and click the green **Code** button. Choose the **Download ZIP** option to download the scripts as a ZIP file.

For more information on MySQL and the tools available, visit *https://dev.mysql.com/doc/*. The MySQL reference manual is particularly helpful. Documentation for MySQL Workbench can be found at *https://dev.mysql.com/doc/workbench/en/*, and for documentation on the MySQL command line you can check out *https://dev.mysql.com/doc/refman/8.0/en/mysql.html*.

MySQL is a fantastic database system to learn. Let's get started!

PART I

GETTING STARTED

In this part of the book, you'll install MySQL and the tools to access your MySQL databases. Then, you'll start getting familiar with these tools by creating your first database.

In Chapter 1, you'll install MySQL, MySQL Workbench, and the MySQL command line client on your computer.

In Chapter 2, you'll create your own MySQL database and a database table.

1

INSTALLING MYSQL AND TOOLS

To begin working with databases, you'll install the free version of MySQL, called *MySQL Community Server* (also known as *MySQL Community Edition*), and two handy tools: MySQL Workbench and the MySQL command line client. This software can be downloaded for free from the MySQL website. You will use these tools to work on projects and exercises later in this book.

The MySQL Architecture

MySQL uses a *client/server architecture*, as shown in Figure 1-1.

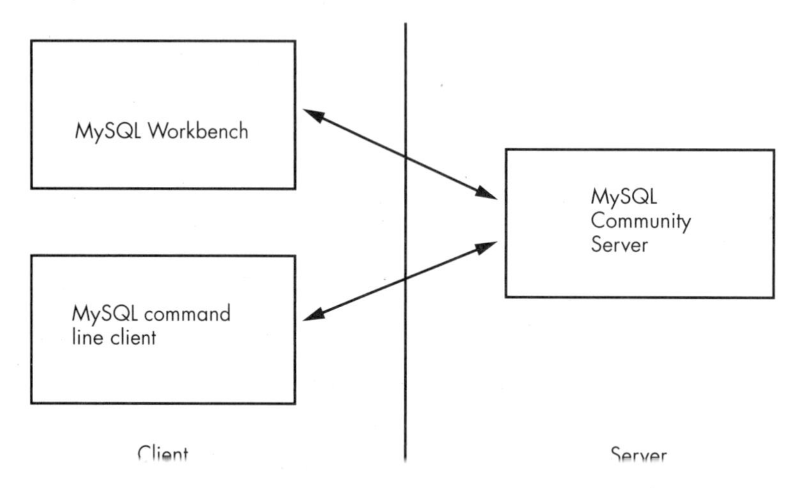

Figure 1-1: The client/server architecture

The server side of this architecture hosts and manages resources or services that the client side needs to access. This means that, in a live production environment, the server software (MySQL Community Server) would run on a dedicated computer housing the MySQL database. The tools used to access the database, MySQL Workbench and the MySQL command line client, would reside on the user's computer.

Because you're setting up a development environment for learning purposes, you'll install both the MySQL client tools and the MySQL Community Server software on the same computer. In other words, your computer will act as both the client and the server.

Installing MySQL

Instructions for installing MySQL are available at *https://dev.mysql.com.* Click **MySQL Documentation**, and under the MySQL Server heading, click **MySQL Reference Manual** and select the most recent version. You'll then be taken to the reference manual for that version. On the left-hand menu, click **Installing and Upgrading MySQL**. Find your operating system in the table of contents and follow the instructions to download and install MySQL Community Server.

There are countless ways to install MySQL—for example, from a ZIP archive, the source code, or a MySQL installer program. The instructions vary based on your operating system and which MySQL products you want to use, so the best and most current resource for installation is always the MySQL website. However, I'll offer a few tips:

- When you install MySQL, it creates a database user called `root` and asks you to choose a password. *Don't lose this password*; you'll need it later.
- In general, I've found it easier to use an installer program, like MySQL Installer, if one is available.
- If you're using Windows, you'll be given the option of two different installers: a web installer or a full bundle installer. However, it's not obvious which one is which, as shown in Figure 1-2.

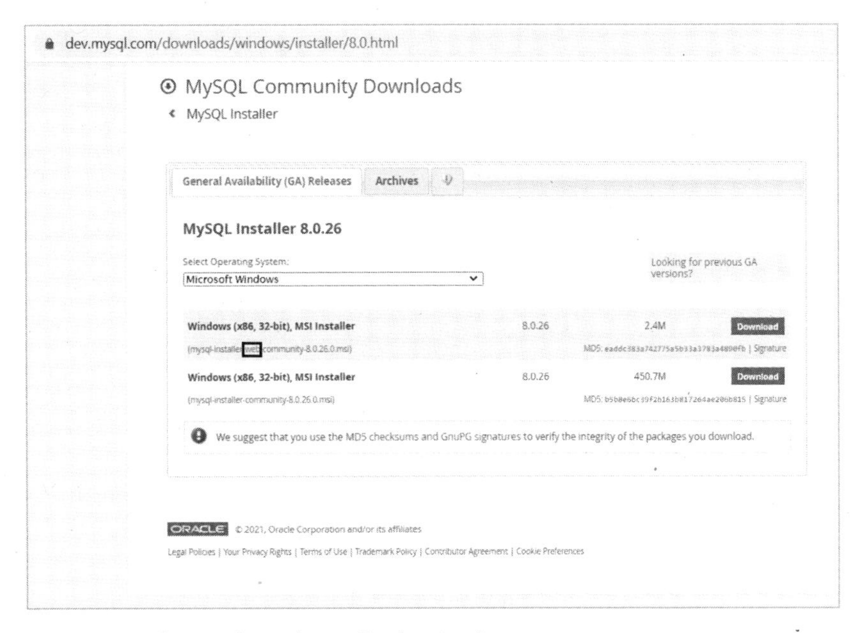

Figure 1-2: Selecting the web installer for Windows

The web installer has a much smaller file size and its filename contains the word *web*, as highlighted in the figure. I recommend choosing this option because it allows you to select the MySQL products you want to install, and it downloads them from the web. The full bundle installer contains all MySQL products, which shouldn't be necessary.

As of this writing, both installers appear on this web page as 32-bit. This refers to the installation application, not MySQL itself. Either installer can install 64-bit binaries. In fact, on Windows, MySQL is available only for 64-bit operating systems.

- You can download MySQL without creating an account if you prefer. On the web page shown in Figure 1-3, select **No Thanks, Just Start My Download** at the bottom of the screen.

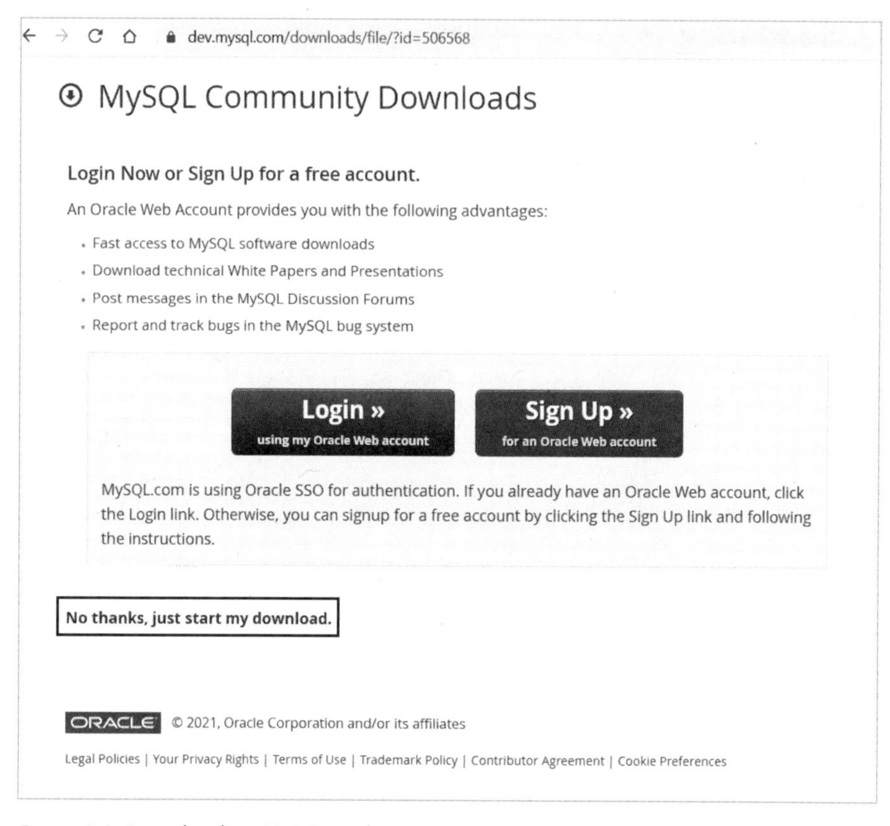

Figure 1-3: Downloading MySQL without creating an account

From here, your next step is to download MySQL Workbench, a graphical tool used to access MySQL databases. With this tool, you can explore your database, run SQL statements against that database, and review the data that gets returned. To download MySQL Workbench, go to *https://dev.mysql.com/doc/workbench/en/*. This takes you directly to the MySQL Workbench reference manual. Click **Installation** in the left-hand menu, choose your operating system, and follow the instructions.

When you install MySQL Community Server or MySQL Workbench on your computer, the MySQL command line client should be installed automatically. This client allows you to connect to a MySQL database from the *command line interface* of your computer (also called the *console*, *command prompt*, or *terminal*). You can use this tool to run a SQL statement, or many SQL statements saved in a script file, against a MySQL database. The MySQL command line client is useful in situations where you don't need to see your results in a nicely formatted graphical user interface.

You'll use these three MySQL products for most of what you do in MySQL, including the exercises in this book.

Now that your computer is set up with MySQL, you can start creating databases!

Summary

In this chapter, you installed MySQL, MySQL Workbench, and the MySQL command line client from the official website. You located the MySQL Server and MySQL Workbench reference manuals, which contain tons of useful information. I recommend using these if you get stuck, have questions, or want to learn more.

In the next chapter, you'll learn how to view and create MySQL databases and tables.

2

CREATING DATABASES AND TABLES

In this chapter, you'll use MySQL Workbench to view and create databases in MySQL. Then you'll learn how to create tables to store data in those databases. You'll define the name of the table and its columns and specify the type of data that the columns can contain. Once you've practiced these basics, you'll improve your tables using two helpful MySQL features, constraints and indexes.

Using MySQL Workbench

As you learned in Chapter 1, MySQL Workbench is a visual tool you can use to enter and run SQL commands and view their results. Here, we'll walk through the basics of how to use MySQL Workbench to view databases.

NOTE *If you're using another tool, such as PhpMyAdmin, MySQL Shell, or the MySQL command line client, be sure to read through this section anyway. The MySQL commands are the same regardless of the tool you use to connect to MySQL.*

You'll start by opening MySQL Workbench by double-clicking its icon. The tool looks like Figure 2-1.

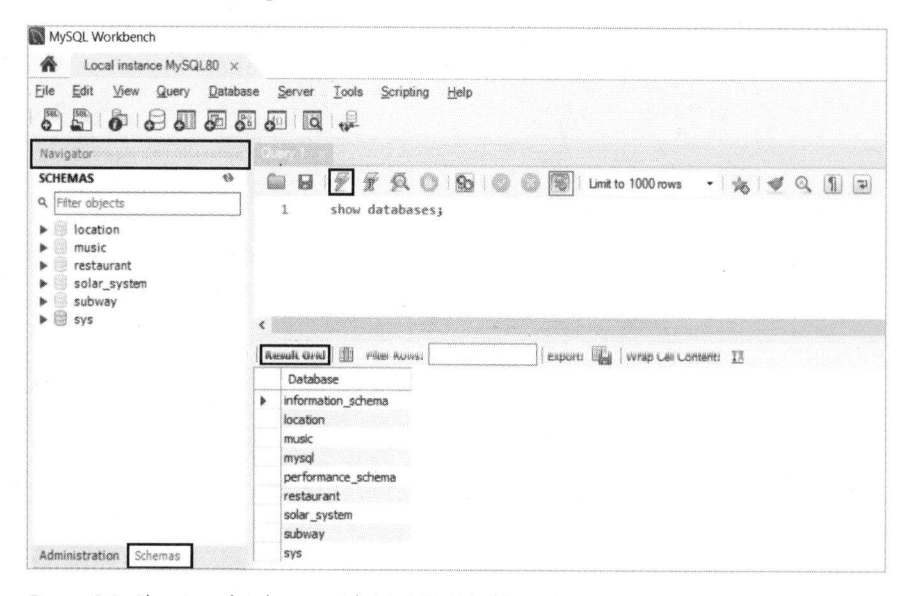

Figure 2-1: Showing databases with MySQL Workbench

In the top-right panel, enter the **`show databases;`** command. Make sure to include the semicolon, which indicates the end of the statement. Then click the lightning bolt icon, highlighted in Figure 2-1, to execute the command. The results, a list of available MySQL databases, appear in the Result Grid panel (your results will look different from mine):

```
Database
--------
information_schema
location
music
mysql
performance_schema
restaurant
solar_system
subway
sys
```

Some databases in this list are system databases that were created automatically when MySQL was installed—such as `information_schema`, `mysql`, and `performance_schema`—and others are databases I've created. Any databases you create should appear in this list.

> **DATABASE TERMINOLOGY**
>
> In MySQL, the term *schema* is synonymous with *database*. You can use `show databases` or `show schemas` to return a list of all of your databases. Notice in Figure 2-1 that *schemas* is the term MySQL Workbench uses.
>
> Sometimes the term *database* gets conflated with *database system* or *relational database management system (RDBMS)*. For example, you might hear somebody ask, "Which database are you using, Oracle, MySQL, or PostgreSQL?" What they're really asking is, "Which database *system* are you using?" A MySQL database is a place to organize your data, while a database system is software that is used to store and retrieve data.

You can also browse databases by using the Navigator panel on the left. Click the **Schemas** tab at the bottom of the panel to show a list of databases, and click the right arrow (▸) to investigate the contents of your databases. Note that, by default, the Navigator panel doesn't show the system databases that were automatically created when MySQL was installed.

Now that you've seen how to view the list of databases in MySQL, it's time to try creating your own.

Creating a New Database

To create a new database, you use the `create database` command along with a name for the database you want to create:

```
create database circus;

create database finance;

create database music;
```

Your database's name should describe the type of data stored there. In this example, the database called `circus` might contain tables for data on clowns, tightrope walkers, and trapeze acts. The `finance` database might have tables for accounts receivable, income, and cash flow. Tables of data on bands, songs, and albums might go in the `music` database.

To remove a database, use the `drop database` command:

```
drop database circus;

drop database finance;

drop database music;
```

These commands remove the three databases you just created, any tables in those databases, and all of the data in those tables.

Of course, you haven't actually created any tables yet. You'll do that now.

> **TRY IT YOURSELF**
>
> **2-1.** Run **show databases** in MySQL Workbench. How many databases do you have?
>
> **2-2.** Using MySQL Workbench, create a database named cryptocurrency. After you have created the database, run the **show databases** command. Do you see the new database in your list of databases? How many databases do you have now?
>
> **2-3.** Drop the cryptocurrency database and then run **show databases** again. Is the database gone from your list of databases?

Creating a New Table

In this example, you'll create a new table to hold global population data and specify what type of data the table can contain:

```
create database land;

use land;

create table continent
(
   continent_id      int,
   continent_name    varchar(20),
   population        bigint
);
```

First, you create a database called land using the create database command you saw earlier. On the next line, the use command tells MySQL to use the land database for the SQL statements that follow it. This ensures that your new table will be created in the land database.

Next, you use the create table command followed by a descriptive name for the table, continent. Within parentheses, you create three columns in the continent table—continent_id, continent_name, and population—and for each column you choose a MySQL data type that controls the type of data allowed in that column. Let's go over this in more detail.

You define the continent_id column as an int so that it will accept integer (numeric) data. Each continent will have its own distinct ID number in this column (1, 2, 3, and so on). Then, you define the continent_name column as a varchar(20) to accept character data up to 20 characters long. Finally, you define the population as a bigint to accept big integers, as the population of an entire continent can be quite a large number.

NOTE *Chapter 4 covers MySQL data types, including* bigint*, in more depth.*

When you run this `create table` statement, MySQL creates an empty table. The table has a table name and its columns are defined, but it doesn't have any rows yet. You can add, delete, and modify the rows in the table whenever you need.

If you try to add a row with data that doesn't match one of the column's data types, however, MySQL will reject the entire row. For example, because the `continent_id` column was defined as an `int`, MySQL won't allow that column to store values like `Continent #1` or `A` because those values contain letters. MySQL won't allow you to store a value like `The Continent of Atlantis` in the `continent_name` column either, since that value has more than 20 characters.

Constraints

When you create your own database tables, MySQL allows you to put *constraints*, or rules, on the data they contain. Once you define constraints, MySQL will enforce them.

Constraints help maintain *data integrity*; that is, they help keep the data in your database accurate and consistent. For example, you might want to add a constraint to the `continent` table so that there can't be two rows in the table with the same value in a particular column.

The constraints available in MySQL are `primary key`, `foreign key`, `not null`, `unique`, `check`, and `default`.

Primary Keys

Identifying the primary key in a table is an essential part of database design. A primary key consists of a column, or more than one column, and uniquely identifies the rows in a table. When you create a database table, you need to determine which column(s) should make up the primary key, because that information will help you retrieve the data later. If you combine data from multiple tables, you'll need to know how many rows to expect from each table and how to join the tables. You don't want duplicate or missing rows in your result sets.

Consider this `customer` table with the columns `customer_id`, `first_name`, `last_name`, and `address`:

```
customer_id  first_name  last_name   address
-----------  ----------  ---------   ---------------------------
     1       Bob         Smith       12 Dreary Lane
     2       Sally       Jones       76 Boulevard Meugler
     3       Karen       Bellyacher  354 Main Street
```

To decide what the primary key for the table should be, you need to identify which column(s) uniquely identifies the rows in the table. For this table, the primary key should be `customer_id`, because every `customer_id` corresponds to only one row in the table.

No matter how many rows might be added to the table in the future, there will never be two rows with the same `customer_id`. This can't be said of any other columns. Multiple people can have the same first name, last name, or address.

A primary key can be composed of more than one column, but even the combination of the first_name, last_name, and address columns isn't guaranteed to uniquely identify the rows. For example, Bob Smith at 12 Dreary Lane might live with his son of the same name.

To designate the customer_id column as the primary key, use the primary key syntax when you create the customer table, as shown in Listing 2-1:

```
create table customer
(
    customer_id     int,
    first_name      varchar(50),
    last_name       varchar(50),
    address         varchar(100),
    primary key (customer_id)
);
```

Listing 2-1: Creating a primary key

Here you define customer_id as a column that accepts integer values and as the primary key for the table.

Making customer_id the primary key benefits you in three ways. First, it prevents duplicate customer IDs from being inserted into the table. If someone using your database tries to add customer_id 3 when that ID already exists, MySQL will give an error message and not insert the row.

Second, making customer_id the primary key prevents users from adding a null value (that is, a missing or unknown value) for the customer_id column. When you define a column as the primary key, it's designated as a special column whose values cannot be null. (You'll learn more about null values later in this chapter.)

Those two benefits fall under the category of data integrity. Once you define this primary key, you can be assured that all rows in the table will have a unique customer_id, and that no customer_id will be null. MySQL will enforce this constraint, which will help keep the data in your database of a high quality.

The third advantage to creating a primary key is that it causes MySQL to create an index. An index will help speed up the performance of SQL queries that select from the table. We'll look at indexes more in the "Indexes" section later in this chapter.

If a table has no obvious primary key, it often makes sense to add a new column that can serve as the primary key (like the customer_id column shown here). For performance reasons, it's best to keep the primary key values as short as possible.

Now let's look at a primary key that consists of more than one column, which is known as a *composite key*. The high_temperature table shown in Listing 2-2 stores cities and their highest temperature by year.

```
city                    year  high_temperature
----------------------  ----  ----------------
Death Valley, CA        2020  130
International Falls, MN 2020  78
```

```
New York, NY              2020   96
Death Valley, CA          2021   128
International Falls, MN   2021   77
New York, NY              2021   98
```

Listing 2-2: Creating multiple primary key columns

For this table, the primary key should consist of both the `city` and `year` columns, because there should be only one row in the table with the same city and year. For example, there's currently a row for Death Valley for the year 2021 with a high temperature of 128, so when you define `city` and `year` as the primary key for this table, MySQL will prevent users from adding a second row for Death Valley for the year 2021.

To make `city` and `year` the primary key for this table, use MySQL's `primary key` syntax with both column names:

```
create table high_temperature
(
    city              varchar(50),
    year              int,
    high_temperature  int,
    primary key (city, year)
);
```

The `city` column is defined to hold up to 50 characters, and the `year` and `high_temperature` columns are defined to hold an integer. The primary key is then defined to be both the `city` and `year` columns.

MySQL doesn't require you to define a primary key for the tables you create, but you should for the data integrity and performance benefits cited earlier. If you can't figure out what the primary key should be for a new table, that probably means you need to rethink your table design.

Every table can have at most one primary key.

Foreign Keys

A foreign key is a column (or columns) in a table that matches the table to the primary key column(s) of another table. Defining a foreign key establishes a relationship between two tables so that you will be able to retrieve one result set containing data from both tables.

You saw in Listing 2-1 that you can create the primary key in the `customer` table using the `primary key` syntax. You'll use similar syntax to create the foreign key in this `complaint` table:

```
create table complaint
    (
    complaint_id  int,
    customer_id   int,
    complaint     varchar(200),
    primary key (complaint_id),
    foreign key (customer_id) references customer(customer_id)
    );
```

In this example, first you create the complaint table, define its columns and their data types, and specify complaint_id as the primary key. Then, the foreign key syntax allows you to define the customer_id column as a foreign key. With the references syntax, you specify that the customer_id column of the complaint table references the customer_id column of the customer table (you'll learn what this means in a moment).

Here's the customer table again:

```
customer_id  first_name  last_name   address
-----------  ----------  ---------   ---------------------------
     1       Bob         Smith       12 Dreary Lane
     2       Sally       Jones       76 Boulevard Meugler
     3       Karen       Bellyacher  354 Main Street
```

And here's the data for the complaint table:

```
complaint_id  customer_id  complaint
------------  -----------  -------------------------------
      1             3      I want to speak to your manager
```

The foreign key allows you to see which customer customer_id 3 in the complaint table is referring to in the customer table; in this case, customer_id 3 references Karen Bellyacher. This arrangement, illustrated in Figure 2-2, allows you to track which customers made which complaints.

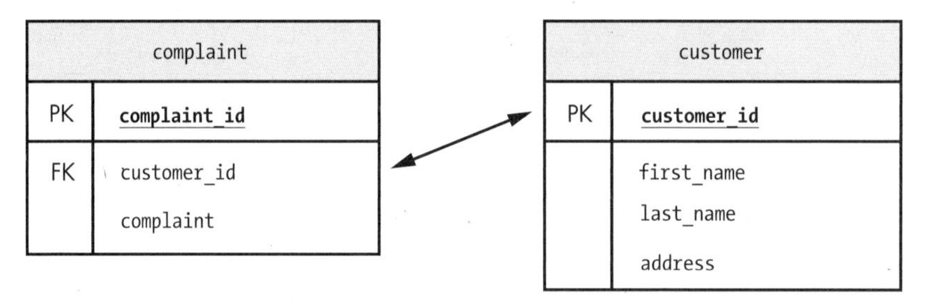

Figure 2-2: Primary keys and foreign keys

In the customer table, the customer_id column has been defined as the primary key (labeled PK). In the complaint table, the customer_id column has been defined as a foreign key (FK) because it will be used to join the complaint table to the customer table.

Here's where things get interesting. Because you defined the foreign key, MySQL won't allow you to add a new row in the complaint table unless it is for a valid customer—that is, unless there is a customer_id row in the customer table that correlates with a customer_id in the complaint table. If you try to add a row in the complaint table for customer_id 4, for example, MySQL will give an error. It doesn't make sense to have a row in the complaint table for a customer that doesn't exist, so MySQL prevents the row from being added in order to maintain data integrity.

Also, now that you've defined the foreign key, MySQL will not allow you to delete customer_id 3 from the customer table. Deleting this ID would leave a row in the complaint table for customer_id 3, which would no longer correspond to any row in the customer table. Restricting data deletion is part of referential integrity.

> **DATA INTEGRITY VS. REFERENTIAL INTEGRITY**
>
> Data integrity refers to the overall accuracy and consistency of the data in your database. You want high-quality data. People lose confidence in your data when they spot problems like a salary value that contains alpha characters or a percent increase value over 100 percent.
>
> Referential integrity refers to the quality of the *relationships* between the data in your tables. If you have a complaint in the complaint table for a customer that doesn't exist in the customer table, you have a referential integrity problem. By defining customer_id as a foreign key, you can be assured that every customer_id in the complaint table refers to a customer_id that exists in the customer table.

There can be only one primary key per table, but a table can have more than one foreign key (see Figure 2-3).

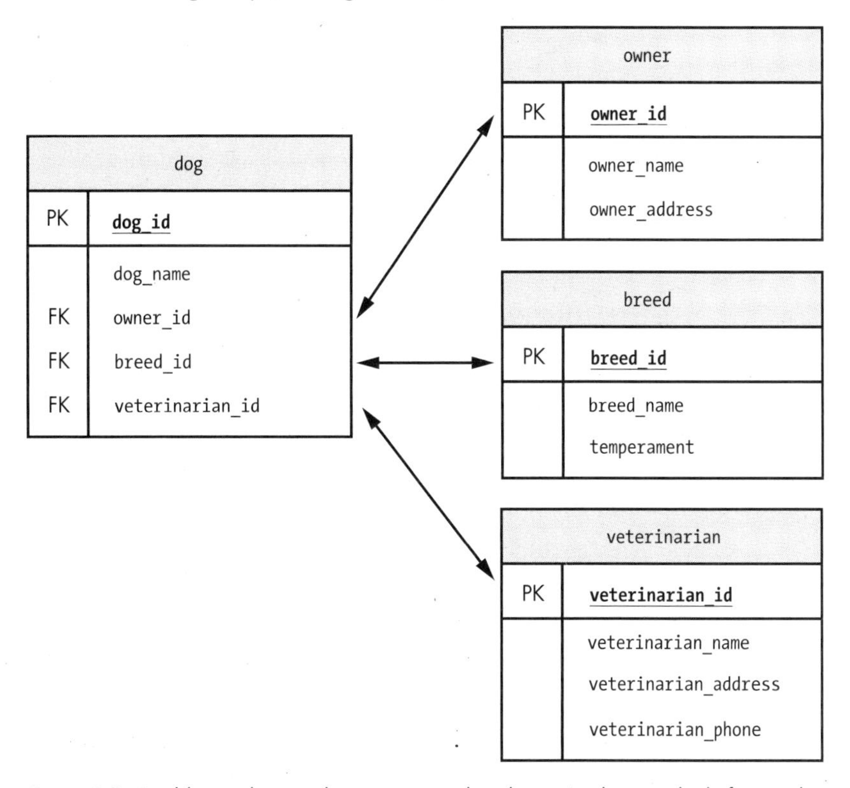

Figure 2-3: A table can have only one primary key, but it can have multiple foreign keys.

Figure 2-3 shows an example of a table named `dog` that has three foreign keys, each pointing to the primary key of a different table. In the `dog` table, `owner_id` is a foreign key used to refer to the `owner` table, `breed_id` is a foreign key used to refer to the `breed` table, and `veterinarian_id` is a foreign key used to refer to the `veterinarian` table.

As with primary keys, when you create a foreign key, MySQL will automatically create an index that will speed up the access to the table. More on that shortly.

TRY IT YOURSELF

2-4. Create a database called `athletic` containing a table called `sport` and a table called `player`. Define the column names, primary key, and foreign key as shown here.

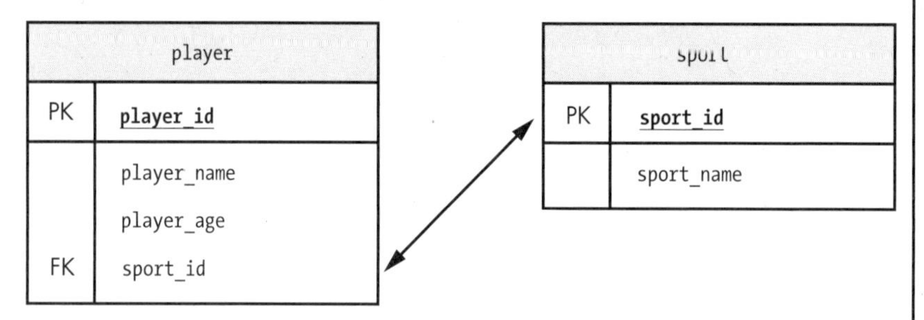

Have all the columns in the tables accept integer values except for `player_name` and `sport_name`, which should accept up to 50 characters. The primary key for the `player` table should be `player_id`. The primary key for the `sport` table should be `sport_id`. The `player` table should have a foreign key of `sport_id` that references the primary key of the `sport` table.

not null

A null value represents an empty or undefined value. It is not the same as zero, an empty character string, or a space character.

Allowing null values in a column can be appropriate in some cases, but other times, permitting the absence of crucial information could result in the database missing data that is needed. Take a look at this table named `contact` that contains contact information:

```
contact_id   name          city       phone      email_address
-----------  ----------    ---------  -------    -----------------
     1       Steve Chen    Beijing    123-3123   steve@schen21.org
     2       Joan Field    New York   321-4321   jfield@jfny99.com
     3       Bill Bashful  Lincoln    null       bb@shyguy77.edu
```

The value of the phone column for contact_id 3 is null because Bill Bashful doesn't own a phone. It is reasonable that the contact table would allow null values for the phone column, as a phone number might not be available or applicable for a contact.

On the other hand, the name column should not allow null values. It would be better not to allow the following row to be added to the contact table:

```
contact_id   name        city       phone    email_address
-----------  ----------  ---------  -------  ---------------
     3       null        Lincoln    null     bb@shyguy77.edu
```

There isn't much point in saving information about a contact unless you know their name, so you can add a not null constraint to the name column to prevent this situation from occurring.

Create the contact table like so:

```
create table contact
(
    contact_id     int,
    name           varchar(50) not null,
    city           varchar(50),
    phone          varchar(20),
    email_address  varchar(50),
    primary key(contact_id)
);
```

Using the not null syntax when you define the name column prevents a value of null from being stored there and maintains data integrity. If you try to add a row with a null name, MySQL will display an error message and the row will be rejected.

For columns defined as the table's primary key, such as the contact_id column in this example, specifying not null isn't necessary. MySQL prevents null values for primary key columns automatically.

unique

If you want to prevent duplicate values in a column, you can add a unique constraint to the column definition. Let's return to the contact table from the previous example:

```
create table contact
(
    contact_id     int,
    name           varchar(50)  not null,
    city           varchar(50),
    phone          varchar(20),
    email_address  varchar(50)  unique,
    primary key(contact_id)
);
```

Here, you prevent duplicate email addresses from being entered by using the `unique` syntax on the `email_address` column. Now MySQL will no longer allow two contacts in the table to have the same email address.

check

You can use a `check` constraint to make sure that a column contains certain values or a certain range of values. For example, let's revisit the `high_temperature` table from Listing 2-2:

```
create table high_temperature
(
    city              varchar(50),
    year              int,
    high_temperature  int,
    constraint check (year between 1880 and 2200),
    constraint check (high_temperature < 200),
    primary key (city, year)
);
```

In this example, you add a `check` constraint to the `year` column to make sure that any year entered into the table is between 1880 and 2200. Accurate temperature tracking wasn't available until 1880, and your database probably won't be in use after the year 2200. Trying to add a year that is outside of that range would most likely be an error, so the constraint will prevent that from occurring.

You've also added a `check` constraint to the `high_temperature` column to limit temperature values to less than 200 degrees, because a temperature higher than that would most likely be a data error.

default

Finally, you can add a `default` constraint to a column so that if a value isn't supplied, a default value will be used. Take a look at the following `job` table:

```
create table job
(
    job_id      int,
    job_desc    varchar(100),
    shift       varchar(50) default '9-5',
    primary key (job_id)
);
```

In this example, you add a `default` constraint to the `shift` column, which stores data on work schedules. The default shift is `9-5`, meaning that if a row doesn't include any data for the shift column, `9-5` will be written to the column. If a value for `shift` is provided, the default won't be used.

You've seen how different constraints can help you improve and maintain the integrity of the data in your tables. Let's turn now to another MySQL feature that also offers performance benefits to your tables: indexes.

Indexes

MySQL lets you create indexes on your tables to speed up the process of retrieving data; in some cases, such as in tables with defined primary or foreign keys, MySQL will create indexes automatically. Just as an index in the back of a book can help you find information without needing to scan each page, indexes help MySQL find data in your tables without having to read every row.

NOTE *You'll see the terms* indexes *and* indices *in MySQL resources and documentation. Both are correct; it's just a matter of individual preference.*

Say you create a product table like so

```
create table product
(
    product_id      int,
    product_name    varchar(100),
    supplier_id     int
);
```

and you want to make the process of retrieving information about suppliers more efficient. Here's the syntax to create an index that will do that:

```
create index product_supplier_index on product(supplier_id);
```

In this example, you create an index, called `product_supplier_index`, on the `supplier_id` column of the `product` table. Now, when users retrieve data from the `product` table using the `supplier_id` column, the index should make that retrieval quicker.

Once you create an index, you won't need to reference it by name—MySQL will use it behind the scenes. The new index won't change anything about the way you use the table; it will just speed up access to it.

Although adding indexes can significantly improve performance, it wouldn't make sense to index every column. Maintaining indexes has a performance cost, and creating indexes that don't get used can actually decrease performance.

When you create tables, MySQL automatically creates most of the indexes that you'll need. You don't need to create indexes for columns that have been defined as primary keys, as foreign keys, or with `unique` constraints, because MySQL automatically indexes those columns.

Let's look at how we would create the `dog` table from Figure 2-3:

```
use pet;

create table dog
(
    dog_id          int,
    dog_name        varchar(50),
    owner_id        int,
    breed_id        int,
```

```
    veterinarian_id   int,
    primary key (dog_id),
    foreign key (owner_id) references owner(owner_id),
    foreign key (breed_id) references breed(breed_id),
    foreign key (veterinarian_id) references veterinarian(veterinarian_id)
);
```

The primary key for the table is `dog_id`, and the foreign keys are `owner_id`, `breed_id`, and `veterinarian_id`. Note that you haven't created any indexes with the `create index` command. MySQL has automatically created indexes, however, from the columns labeled as the primary key and the foreign keys. You can confirm this using the `show indexes` command:

```
show indexes from dog;
```

The results are shown in Figure 2-4.

Table	Non_unique	Key_name	Seq_in_index	Column_name	Collation	Cardinality	Sub_part	Packed	Null	Index_type	Comment	Index_comment	Visible	Expression
dog	0	PRIMARY	1	dog_id	A	0	NULL	NULL		BTREE			YES	NULL
dog	1	owner_id	1	owner_id	A	0	NULL	NULL	YES	BTREE			YES	NULL
dog	1	breed_id	1	breed_id	A	0	NULL	NULL	YES	BTREE			YES	NULL
dog	1	veterinarian_id	1	veterinarian_id	A	0	NULL	NULL	YES	BTREE			YES	NULL

Figure 2-4: Indexes automatically created by MySQL for the dog table

You can see in the `Column_name` column that MySQL automatically created all of the indexes that you need for this table.

The `owner`, `breed`, *and* `veterinarian` *tables must exist before the dog table gets created. The code to create those tables in the pet database is in* https://github.com/ricksilva/mysql_cc/blob/main/chapter_2.sql.

Dropping and Altering Tables

To *drop* a table, which removes the table and all of its data, use the `drop table` syntax:

```
drop table product;
```

Here you tell MySQL to drop the `product` table you created in the previous section.

To make changes to a table, use the `alter table` command. You can add columns, drop columns, change a column's data type, rename columns, rename the table, add or remove constraints, and make other changes.

Try altering the `customer` table from Listing 2-1:

```
alter table customer add column zip varchar(50);
alter table customer drop column address;
alter table customer rename column zip to zip_code;
alter table customer rename to valued_customer;
```

Here you alter the `customer` table in four ways: you add a column named `zip` that stores zip codes, remove the `address` column, rename the `zip` column to `zip_code` to make it more descriptive, and change the table name from `customer` to `valued_customer`.

WARNING *If you drop a table, you'll lose all the data in the table as well.*

Summary

In this chapter, you saw how to use MySQL Workbench to run commands and view databases. You created your own database tables and learned how to optimize them using indexes and adding constraints.

In the next chapter, the beginning of Part II of the book, you'll learn about retrieving data from MySQL tables using SQL, displaying your data in an ordered way, formatting SQL statements, and using comments in SQL.

PART II

SELECTING DATA FROM A MYSQL DATABASE

So far, you've created MySQL databases and tables for storing data. In Part II, you'll retrieve data from those tables.

In Chapter 3, you'll select data from a MySQL table.

In Chapter 4, you'll look more into MySQL data types.

In Chapter 5, you'll use joins to select data from multiple tables.

In Chapter 6, you'll dig deeper into complex joins with multiple tables.

In Chapter 7, you'll learn more about comparing values in MySQL.

In Chapter 8, you'll learn about MySQL's built-in functions and call them from your SQL statements.

3

INTRODUCTION TO SQL

To select data from a MySQL database, you'll use *Structured Query Language (SQL)*. SQL is the standard language for querying and managing data in an RDBMS like MySQL.

SQL commands can be categorized into *Data Definition Language (DDL)* statements and *Data Manipulation Language (DML)* statements. So far, you've been using DDL commands like `create database`, `create table`, and `drop table` to *define* your databases and tables.

DML commands, on the other hand, are used to *manipulate* the data in your existing databases and tables. In this chapter, you'll use the DML `select` command to retrieve data from a table. You'll also learn how to specify an order for MySQL to sort your results and how to deal with null values in your table columns.

NOTE *Some people pronounce SQL as "sequel" and others say "ess-cue-ell." Whichever way you like to pronounce it, SQL is the main language used in MySQL development, so it pays to learn it well.*

Querying Data from a Table

A *query* is a request for information from a database table or group of tables. To specify the information you want to retrieve from the table, use the select command, as shown in Listing 3-1.

```
select continent_id,
       continent_name,
       population
from   continent;
```

Listing 3-1: Using select to display data from the continent table

Here you're querying the continent table (as indicated by the from keyword), which contains information about each continent's name and population. Using the select command, you specify that you want to return data from the continent_id, continent_name, and population columns. This is known as a select statement.

Listing 3-2 shows the results of running the select statement.

```
continent_id  continent_name  population
------------  --------------  ----------
       1      Asia            4641054775
       2      Africa          1340598147
       3      Europe           747636026
       4      North America    592072212
       5      South America    430759766
       6      Australia         43111704
       7      Antarctica               0
```

Listing 3-2: Results of running the select statement

The query returned a list of all seven continents, displaying each continent's ID, name, and population.

In order to show the data from only one continent—Asia, for example—you can add a where clause to the end of your previous code:

```
select continent_id,
       continent_name,
       population
from   continent
where  continent_name = 'Asia';
```

A where clause filters the results by applying conditions to the select statement. This query finds the only row in the table where the value of the continent_name column equals Asia and displays the following result:

```
continent_id  continent_name  population
------------  --------------  ----------
       1      Asia            4641054775
```

Now change the select statement to select only the population column:

```
select population
from   continent
where  continent_name = 'Asia';
```

The query now returns one column (population) for one row (Asia):

```
population
----------
4641054775
```

The continent_id and continent_name values don't appear in your result set because you didn't select them in the SQL query.

> **TRY IT YOURSELF**
>
> Locate the commands in the SQL script at *https://github.com/ricksilva/mysql_cc/blob/main/chapter_3.sql* that create the feedback database and that create and load the customer table. Copy the SQL commands from the scripts and paste them into MySQL Workbench. Run them to create the database and the table, and to load its rows.
>
> **3-1.** The feedback database contains a table called customer. Select the first_name and last_name columns from the table for all customers.
>
> **3-2.** Modify your query to select the customer_id, first_name, and last_name columns from the customer table for all customers whose first name is Karen. How many Karens do you have in the table?

Using the Wildcard Character

The asterisk wildcard character (*) in SQL allows you to select all of the columns in a table without having to type all of their names in your query:

```
select *
from   continent;
```

This query returns all three columns from the continent table. The results are the same as those for Listing 3-1, where you individually listed the three column names.

Ordering Rows

When you query data from your database, you'll often want to see the results in a particular order. To do that, add an order by clause to your SQL query:

```
select continent_id,
       continent_name,
```

```
       population
from   continent
order by continent_name;
```

Here you select all of the columns in the continent table and order the results alphabetically by the values in the continent_name column.

The results are as follows:

```
continent_id  continent_name  population
------------  --------------  ----------
       2      Africa          1340598147
       7      Antarctica               0
       1      Asia            4641054775
       6      Australia         43111704
       3      Europe           747636026
       4      North America    592072212
       5      South America    430759766
```

Adding order by continent_name results in an alphabetized list, regardless of the values of the continent_id or population columns. MySQL ordered the rows alphabetically because continent_name is defined as a column that stores alphanumeric characters.

CHARACTER SETS AND COLLATIONS

Interestingly, the characters that can be stored and the order in which they get sorted may be different in your MySQL environment than in mine. This is determined by the character set and collation you're using.

A *character set* defines the set of characters that can be stored. *Collations* are the rules for comparing character sets. Examples of character sets are latin1, utf8mb3, and utf8mb4. The default character set, as of this writing, is utf8mb4. It allows you to save a wide range of characters and even use emojis in your text columns.

The default collation is utf8mb4_0900_ai_ci. The *_ci* stands for "case insensitive." There is also a utf8mb4_0900_ai_cs collation, where the *_cs* means "case sensitive." If you are using the case-insensitive collation but I have switched to the case-sensitive collation, our results will be sorted differently.

MySQL can also order columns with integer data types. You can specify whether you want your results sorted in ascending (lowest to highest) or descending (highest to lowest) order using the asc and desc keywords:

```
select continent_id,
       continent_name,
       population
from   continent;
order by population desc;
```

In this example, you have MySQL order the results by `population` and sort the values in descending order (`desc`) order.

NOTE *If you don't specify* `asc` *or* `desc` *in your* `order by` *clause for integer data types, MySQL will default to ascending.*

The results are as follows:

```
continent_id  continent_name  population
------------  --------------  ----------
      1       Asia            4641054775
      2       Africa          1340598147
      3       Europe           747636026
      4       North America    592072212
      5       South America    430759766
      6       Australia         43111704
      7       Antarctica               0
```

The query returns all seven rows because you didn't filter the results with a `where` clause. Now the data is displayed in descending order based on the `population` column instead of alphabetically based on the `continent_name` column.

Formatting SQL Code

So far, the SQL you've seen has been in a nice, readable format:

```
select continent_id,
       continent_name,
       population
from   continent;
```

Notice how the column names and the table name all align vertically. It's a good idea to write SQL statements in a neat, maintainable format like this, but MySQL will also allow you to write SQL statements in less organized ways. For example, you can write the code from Listing 3-1 on only one line:

```
select continent_id, continent_name, population from continent;
```

Or you can separate the `select` and `from` statements, like so:

```
select continent_id, continent_name, population
from continent;
```

Both options return the same results as Listing 3-1, though your SQL might be a little harder for people to understand.

Readable code is important for the maintainability of your codebase, even though MySQL will run less readable code without issue. It might be tempting to just get the code working and then move on to the next task, but writing the code is only the first part of your job. Take the time to make your code readable, and your future self (or whoever will be maintaining the code) will thank you.

Let's look at some other SQL code conventions you might see.

Uppercase Keywords

Some developers use uppercase for MySQL keywords. For example, they might write Listing 3-1 like this, with the words select and from in uppercase:

```
SELECT continent_id,
       continent_name,
       population
FROM   continent;
```

Similarly, some developers might format a create table statement with multiple phrases in uppercase:

```
CREATE TABLE dog
(
    dog_id             int,
    dog_name           varchar(50) UNIQUE,
    owner_id           int,
    breed_id           int,
    veterinarian_id    int,
    PRIMARY KEY (dog_id),
    FOREIGN KEY (owner_id) REFERENCES owner(owner_id),
    FOREIGN KEY (breed_id) REFERENCES breed(breed_id),
    FOREIGN KEY (veterinarian_id) REFERENCES veterinarian(veterinarian_id)
);
```

Here, create table, unique, primary key, foreign key, and references have all been capitalized for readability. Some MySQL developers would capitalize the data types int and varchar as well. If you find using uppercase for keywords is beneficial, feel free to do so.

If you are working with an existing codebase, it's best to be consistent and follow the coding style precedent that has been set. If you work at a company that has formal style conventions, you should follow them. Otherwise, choose what works best for you. You'll get the same results either way.

Backticks

If you maintain SQL that other developers have written, you may encounter SQL statements that use backticks (`):

```
select `continent_id`,
       `continent_name`,
       `population`
from   `continent`;
```

This query selects all of the columns in the continent table, surrounding the column names and the table name with backticks. In this example, the statement runs just as well without the backticks.

Backticks allow you to get around some of MySQL's rules for naming tables and columns. For example, you might have noticed that when column names consist of more than one word, I've used an underscore between the words instead of a space, like `continent_id`. If you wrap column names in backticks, however, you don't need to use underscores; you can name a column `continent id` rather than `continent_id`.

Normally, if you were to name a table or column `select`, you'd receive an error message because `select` is a MySQL *reserved word*; that is, it has a dedicated meaning in SQL. However, if you wrap `select` in backticks, the query will run without error:

```
select * from `select`;
```

In this `select * from` statement, you're selecting all columns within the `select` table.

Although MySQL will run code like this, I recommend avoiding backticks, as your code will be more maintainable and easier to type without them. In the future, another developer who needs to make a change to this query might be confused by a table named `select` or a table with spaces in its name. Your goal should always be to write code that is simple and well organized.

Code Comments

Comments are lines of explanatory text that you can add to your code to make it easier to understand. They can help you or other developers maintain the code in the future. Oftentimes, comments clarify complex SQL statements or point out anything about the table or data that's out of the ordinary.

To add single-line comments, use two hyphens followed by a space (--). This syntax tells MySQL that the rest of the line is a comment.

This SQL query includes a one-line comment at the top:

```
-- This SQL statement shows the highest-populated continents at the top
select continent_id,
       continent_name,
       population
from   continent
order by population desc;
```

You can use the same syntax to add a comment at the end of a line of SQL:

```
select continent_id,
       continent_name, -- Continent names are displayed in English
       population
from   continent
order by population desc;
```

In this code, the comment for the `continent_name` column lets developers know that the names are displayed in English.

WHITESPACE IN SINGLE-LINE COMMENTS

At first blush, it seems strange that MySQL will allow this comment:

```
select 7+5; -- Adding 5 dollars to the 7 I already had
```

but not this one:

```
select 7+5; --Adding 5 dollars to the 7 I already had
```

Why does MySQL require whitespace after the --?

The answer is that a comment using the -- syntax could be confused with the syntax for subtracting a negative number. To subtract –5 from 7 in SQL, you could use this syntax:

```
select 7--5;
```

If MySQL always interpreted -- (two hyphens with no spaces) as a comment, then this line of code would return 7, not the correct result of 12, because everything after the 7 would get commented out.

To add multiline comments, use /* at the beginning of the comment and */ at the end:

```
/*
This query retrieves data for all the continents in the world.
The population of each continent is updated in this table yearly.
*/
select * from continent;
```

This two-line comment explains the query and says how often the table is updated.

The syntax for inline comments is similar:

```
select 3.14 /* The value of pi */ * 81;
```

There are some special uses for inline comments. For example, if you maintain code that has been written by others, you might notice what looks like cryptic inline comments:

```
select /*+ no_index(employee idx1) */
       employee_name
from   employee;
```

The `/*+ no_index(employee idx1) */` in the first line is an *optimizer hint*, which uses the inline comment syntax with a plus sign after the /*.

When you run a query, MySQL's query optimizer tries to determine the fastest way to execute it. For example, if there are indexes on the `employee` table, would it be faster to use the indexes to access the data, or do the tables have so few rows that using the indexes would actually be slower?

The query optimizer usually does a good job of coming up with query plans, comparing them, and then executing the fastest plan. But there are times when you'll want to give your own instructions—hints—about the most efficient way to execute the query.

The hint in the preceding example tells the optimizer not to use the `idx1` index on the `employee` table.

Query optimization is a vast topic and we've barely scratched the surface, but if you encounter the /*+ . . . */ syntax, just know that it allows you to provide hints to MySQL.

As you can see, a well-placed, descriptive comment will save time and aggravation. A quick explanation about why you used a particular approach can spare another developer from having to research the same issue, or jog your own memory if you'll be the one maintaining the code. However, avoid the temptation to add comments that state the obvious; if a comment won't make the SQL more understandable, you shouldn't add it. Also, it's important to update comments as you update your code. Comments that aren't up to date and are no longer relevant don't serve a purpose, and might confuse other developers or your future self.

Null Values

As discussed in Chapter 2, `null` represents a missing or unknown value. MySQL has special syntax, including `is null` and `is not null`, to help handle null values in your data.

Consider a table called `unemployed` that has two columns: `region_id` and `unemployed`. Each row represents a region and tells you how many people are unemployed in that region. Look at the full table using `select * from`, like so:

```
select *
from   unemployed;
```

The results are as follows:

```
region_id   unemployed
---------   ----------
    1          2218457
    2           137455
    3             null
```

Regions 1 and 2 have reported their number of unemployed people, but region 3 hasn't done so yet, so the `unemployed` column for region 3 is set to the `null` value. You wouldn't want to use `0` here, because that would mean there are no unemployed people in region 3.

To show only the rows for regions that have an `unemployed` value of `null`, use the `where` clause with `is null`:

```
select *
from   unemployed
where  unemployed is null;
```

The result is:

```
region   unemployed
------   ----------
   3           null
```

On the other hand, if you wanted to *exclude* rows that have an `unemployed` value of `null` in order to see only the data that has already been reported, replace `is null` with `is not null` in the `where` clause, like so:

```
select *
from   unemployed
where  unemployed is not null;
```

The results are as follows:

```
region   unemployed
------   ----------
   1        2218457
   2         137455
```

Using this syntax with null values can help you filter your table data so that MySQL returns only the most meaningful results.

Summary

In this chapter, you learned how to use the `select` statement and the wildcard character to retrieve data from a table, and you saw that MySQL can return results in an order you specify. You also looked at ways to format your code for readability and clarity, including adding comments to your SQL statements to make maintaining the code easier. Finally, you saw how you might handle null values in your data.

Chapter 4 is all about MySQL data types. So far, the tables you've created have mainly used `int` to accept integer data or `varchar` to accept character data. Next, you'll learn more about other MySQL data types for numeric and character data, as well as data types for dates and very large values.

4

MYSQL DATA TYPES

In this chapter, you'll look at all of the available MySQL data types. You've already seen that `int` and `varchar` can be used for integer and character data, but MySQL also has data types to store dates, times, and even binary data. You'll explore how to choose the best data types for your columns and the pros and cons of each type.

When you create a table, you define each column's data type based on the kind of data you'll store in that column. For example, you wouldn't use a data type that allows only numbers for a column that stores names. You might additionally consider the range of values that the column will have to accommodate. If a column needs to store a value like 3.1415, you should use a data type that allows decimal values with four positions after the decimal point. Lastly, if more than one data type can handle the values your column will need to store, you should choose the one that uses the least amount of storage.

Say you want to create a table, `solar_eclipse`, that includes data about solar eclipses, including the date of the eclipse, the time it occurs, the type of eclipse, and its magnitude. Your raw data might look like Table 4-1.

Table 4-1: Data on Solar Eclipses

Eclipse date	Time of greatest eclipse	Eclipse type	Magnitude
2022-04-30	20:42:36	Partial	0.640
2022-10-25	11:01:20	Partial	0.862
2023-04-20	04:17:56	Hybrid	1.013

In order to store this data in a MySQL database, you'll create a table with four columns:

```
create table solar_eclipse
(
    eclipse_date                date,
    time_of_greatest_eclipse    time,
    eclipse_type                varchar(10),
    magnitude                   decimal(4,3)
);
```

In this table, each of the four columns has been defined with a different data type. Since the `eclipse_date` column will store dates, you use the `date` data type. The `time` data type, which is designed to store time data, is applied to the `time_of_greatest_eclipse` column.

For the `eclipse_type` column, you use the `varchar` data type because you need to store variable-length character data. You don't expect these values to be long, so you use `varchar(10)` to set the maximum number of characters to 10.

For the `magnitude` column, you use the `decimal` data type and specify that the values will have four digits total and three digits after the decimal point.

Let's look at these and several other data types in more depth, and explore when it's appropriate to use each one.

String Data Types

A *string* is a set of characters, including letters, numbers, whitespace characters like spaces and tabs, and symbols like punctuation marks. For values that include only numbers, you should use a numeric data type rather than a string data type. You would use a string data type for a value like `I love MySQL 8.0!` but a numeric data type for a value like `8.0`.

This section will examine MySQL's string data types.

char

The `char` data type is used for *fixed-length* strings—that is, strings that hold an exact number of characters. To define a column within a `country_code` table to store three-letter country codes like `USA`, `GBR`, and `JPN`, use `char(3)`, like so:

```
create table country_code
(
    country_code    char(3)
);
```

When defining columns with the char data type, you specify the length of the string inside the parentheses. The char data type defaults to one character if you leave out the parentheses, though in cases where you want only one character, it's clearer to specify char(1) than just char.

The length of the string cannot exceed the length defined within the parentheses. If you tried to insert JAPAN into the country_code column, MySQL would reject the value because the column has been defined to store a maximum of three characters. However, MySQL will allow you to insert a string with fewer than three characters, such as JP; it simply adds a space to the end of JP and saves the value in the column.

You can define a char data type with up to 255 characters. If you try to define a column with a data type of char(256) you'll get an error message because it's out of char's range.

varchar

The varchar data type, which you've seen before, is for *variable-length* strings, or strings that can hold *up to* a specified number of characters. It's useful when you need to store strings but aren't sure exactly how long they will be. For example, to create an interesting_people table and then define a column called interesting_name that stores various names, you need to be able to accommodate short names like Jet Li as well as long names like Hubert Blaine Wolfeschlegelsteinhausenbergerdorff:

```
create table interesting_people
(
    interesting_name    varchar(100)
);
```

In the parentheses, you define a character limit of 100 for the interesting_name column because you don't anticipate that anybody's name in the database will be over 100 characters.

The number of characters that varchar can accept depends on your MySQL configuration. Your database administrator (DBA) can help you, or you can use this quick hack to determine your maximum. Write a create table statement with a column that has an absurdly long varchar maximum value:

```
create table test_varchar_size
(
    huge_column varchar(999999999)
);
```

The create table statement will fail, giving you an error message like

```
Error Code: 1074. Column length too big for column 'huge_column'
(max = 16383);
use BLOB or TEXT instead
```

The table was not created because the varchar definition was too large, but the error message told you that the maximum number of characters that varchar can accept in this environment is 16,383, or varchar(16383).

The varchar data type is mostly used for small strings. When you're storing more than 5,000 characters, I recommend using the text data type instead (we'll get to it momentarily).

enum

The enum data type, short for *enumeration*, lets you create a list of values that you want to allow in a string column. Here's how to create a table called student with a student_class column that can accept only one of the following values—Freshman, Sophomore, Junior, or Senior:

```
create table student
    (
    student_id     int,
    student_class  enum('Freshman','Sophomore','Junior','Senior')
    );
```

If you try to add a value to the column other than the ones in the list of permitted values, it will be rejected. You can add only one of the permitted values to the student_class column; a student can't be both a freshman and a sophomore.

set

The set data type is similar to the enum data type, but set allows you to select multiple values. In the following create table statement, you define a list of languages for a language_spoken column in a table called interpreter:

```
create table interpreter
    (
    interpreter_id     int,
    language_spoken    set('English','German','French','Spanish')
    );
```

The set data type allows you to add any or all of the languages in the set to the language_spoken column, as someone might speak one or more of these languages. If you try to add any value to the column other than the ones in the list, however, they will be rejected.

tinytext, text, mediumtext, and longtext

MySQL includes four text data types that store variable-length strings:

tinytext	Stores up to 255 characters
text	Stores up to 65,535 characters, which is approximately 64KB
mediumtext	Stores up to 16,777,215 characters, approximately 16MB
longtext	Stores up to 4,294,967,295 characters, approximately 4GB

The following create table statement creates a table named book that contains four columns. The last three columns, author_bio, book_proposal, and entire_book, all use text data types of different sizes:

```
create table book
    (
    book_id            int,
    author_bio         tinytext,
    book_proposal      text,
    entire_book        mediumtext
    );
```

You use the tinytext data type for the author_bio column because you don't anticipate any author biographies larger than 255 characters. This also forces users to make sure their bios have fewer than 255 characters. You choose the text data type for the book_proposal column because you aren't expecting any book proposals of over 64KB. Finally, you choose the mediumtext data type for the entire_book column to limit the size of books to 16MB.

STRING FORMATTING

String values must be surrounded by single quotes or double quotes. The following query uses single quotes around the string Town Supply:

```
select  *
from    store
where   store_name = 'Town Supply';
```

This query uses double quotes:

```
select  *
from    store
where   store_name = "Town Supply";
```

Both queries return the same values. Things get more interesting when you want to compare strings that have special characters in them, like apostrophes, quotes, or tabs. For example, using single quotes for Town Supply works fine, but using single quotes for the string Bill's Supply

```
select  *
from    store
where   store_name = 'Bill's Supply';
```

results in the following error:

```
Error Code: 1064. You have an error in your SQL syntax; check the manual
that corresponds to your MySQL server version for the right syntax to use
near 's Supply'' at line 1
```

(continued)

MySQL is confused because the single quotes at the beginning and the end of the string are the same character as the apostrophe in `Bill's`. It's not clear whether the apostrophe is ending the string or part of the string.

You can work around the problem by surrounding the string in double quotes instead of single quotes:

```
select  *
from    store
where   store_name = "Bill's Supply";
```

Now MySQL knows that the apostrophe is part of the string.

You can also fix the error by surrounding the string in single quotes and *escaping* the apostrophe:

```
select  *
from    store
where   store_name = 'Bill\'s Supply';
```

The backslash character is the escape character, and it creates an *escape sequence* that tells MySQL the next character is part of the string. There are other escape sequences available as well:

`\"` Double quote

`\n` Newline (linefeed)

`\r` Carriage return

`\t` Tab

`\\` Backslash

You can use escape sequences to add special characters to strings, like the double quotes around the nickname Kitty:

```
select  *
from    accountant
where   accountant_name = "Kathy \"Kitty\" McGillicuddy";
```

In this case you could also wrap the string in single quotes so you don't have to escape the double quotes:

```
select  *
from    accountant
where   accountant_name = 'Kathy "Kitty" McGillicuddy';
```

Either way, the result returned is `Kathy "Kitty" McGillicuddy`.

Binary Data Types

MySQL provides data types to store *binary* data, or raw data in byte format that is not human-readable.

tinyblob, blob, mediumblob, and longblob

A *binary large object (BLOB)* is a variable-length string of bytes. You can use BLOBs to store binary data like images, PDF files, and videos. BLOB data types come in the same sizes as the text data types. While tinytext can store up to 255 characters, tinyblob can store up to 255 bytes.

tinyblob	Stores up to 255 bytes
blob	Stores up to 65,535 bytes, approximately 64KB
mediumblob	Stores up to 16,777,215 bytes, approximately 16MB
longblob	Stores up to 4,294,967,295 bytes, approximately 4GB

binary

The binary data type is for fixed-length binary data. It's similar to the char data type, except that it's used for strings of binary data rather than character strings. You specify the size of the byte string within the parentheses like so:

```
create table encryption
    (
    key_id          int,
    encryption_key  binary(50)
    );
```

For the column called encryption_key in the encryption table, you set the maximum size of the byte string to 50 bytes.

varbinary

The varbinary data type is for variable-length binary data. You specify the maximum size of the byte string within the parentheses:

```
create table signature
    (
    signature_id    int,
    signature       varbinary(400)
    );
```

Here, you're creating a column called signature (in a table of the same name) with a maximum size of 400 bytes.

bit

One of the lesser-used data types, `bit` is used for storing bit values. You can specify how many bits you want to store, up to a maximum of 64. A definition of `bit(15)` allows you to store up to 15 bits.

STRINGS OF CHARACTERS VS. STRINGS OF BYTES

A *character string*, usually just called a *string*, is a human-readable set of characters. A *byte string*, on the other hand, is a string of bytes. Byte strings aren't human-readable.

In the following table, named `animal`, the `animal_desc` column has been defined with the `tinytext` data type and the `animal_picture` column has been defined with the `mediumblob` data type:

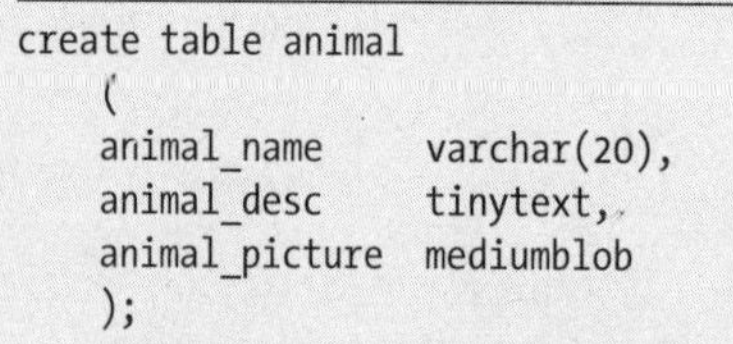

```
create table animal
    (
    animal_name     varchar(20),
    animal_desc     tinytext,
    animal_picture  mediumblob
    );
```

Here is the result when you query the table using MySQL Workbench.

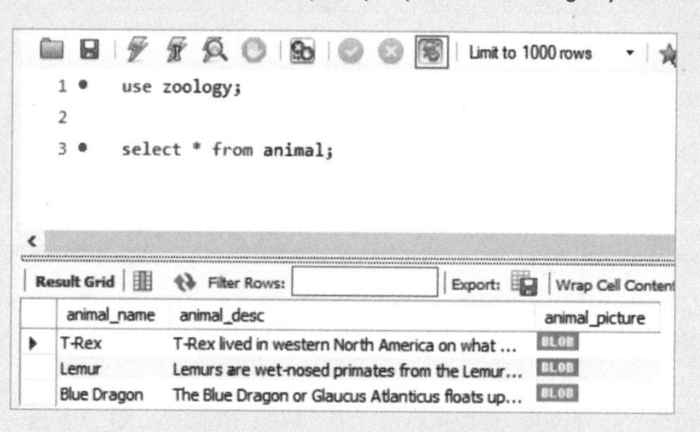

The contents of the `animal_desc` column are human-readable, but MySQL Workbench displays the contents of `animal_picture` as `BLOB` because that column value is set to a byte string format.

Numeric Data Types

MySQL provides data types to store numbers of different sizes. The numeric type to use also depends upon whether the numbers you want to store contain decimal points.

tinyint, smallint, mediumint, int, and bigint

Integers are whole numbers without a fraction or decimal. Integer values can be positive, negative, or zero. MySQL includes the following integer data types:

tinyint Stores integer values that range from –128 to 127, or 1 byte of storage

smallint Stores integer values ranging from –32,768 to 32,767, or 2 bytes of storage

mediumint Stores integer values ranging from –8,388,608 to 8,388,607, or 3 bytes of storage

int Stores integer values from –2,147,483,648 to 2,147,483,647, or 4 bytes of storage

bigint Stores integer values that range from –9,223,372,036,854,775,808 to 9,223,372,036,854,775,807, or 8 bytes of storage

How do you know which integer type is right for your data? Take a look at the `planet_stat` table in Listing 4-1.

```
create table planet_stat
(
    planet              varchar(20),
    miles_from_earth    bigint,
    diameter_km         mediumint
);
```

Listing 4-1: Creating a table on planet statistics

This table contains statistics about planets using `varchar(20)` to store the planet's name, `bigint` to store its distance from Earth in miles, and `mediumint` for the planet's diameter (in kilometers).

Looking at the results, you can see that Neptune is 2,703,959,966 miles from Earth. In this case, `bigint` is the appropriate choice for that column, as `int` wouldn't have been large enough for that value.

```
planet   miles_from_earth  diameter_km
-------  ----------------  -----------
Mars             48678219         6792
Jupiter         390674712       142984
Saturn          792248279       120536
Uranus         1692662533        51118
Neptune        2703959966        49528
```

Considering that `int` takes 4 bytes of storage and `bigint` takes 8 bytes, using `bigint` for a column where `int` would have been large enough means taking up more disk space than necessary. In small tables, using `int` where a `smallint` or a `mediumint` would have sufficed won't cause any problems. But if your table has 20 million rows, it pays to take the time to size the columns correctly—those extra bytes add up.

One technique you can use for space efficiency is defining integer data types as unsigned. By default, the integer data types allow you to store negative and positive integers. If you won't need any negative numbers, you can use unsigned to prevent negative values and increase the number of positive numbers. For example, the tinyint data type gives you a default range of values between –128 and 127, but if you specify unsigned, your range becomes 0 to 255.

If you specify smallint as unsigned, your range becomes 0 to 65,535. Specifying the mediumint data type gives you a range of 0 to 16,777,215, and specifying int changes the range to 0 through 4,294,967,295.

In Listing 4-1, you defined the miles_from_earth column as a bigint, but if you take advantage of the larger unsigned upper range values, you can fit the values into an int data type instead. You can be confident using unsigned for this column, as it will never need to store a negative number—no planet will ever be less than zero miles away from Earth:

```
create table planet_stat
(
    planet            varchar(20),
    miles_from_earth  int unsigned, -- Now using int unsigned, not bigint
    diameter_km       mediumint
);
```

By defining the column as unsigned, you can use the more compact int type and save disk space.

Boolean

Boolean values have only two states: true or false; on or off; 1 or 0. Technically, MySQL doesn't have a data type to capture boolean values; they're stored in MySQL as tinyint(1). You can use the synonym bool to create columns to store boolean values. When you define a column as bool, it creates a tinyint(1) column behind the scenes.

This table called food has two boolean columns, organic_flag and gluten_free_flag, to tell you whether a food is organic or gluten-free:

```
create table food
(
    food              varchar(30),
    organic_flag      bool,
    gluten_free_flag  bool
);
```

It's common practice to add the suffix _flag to columns that contain boolean values, such as organic_flag, because setting the value to true or false can be compared to raising or lowering a flag, respectively.

To view the structure of a table, you can use the describe, or desc, command. Figure 4-1 shows the result of running desc food; in MySQL Workbench.

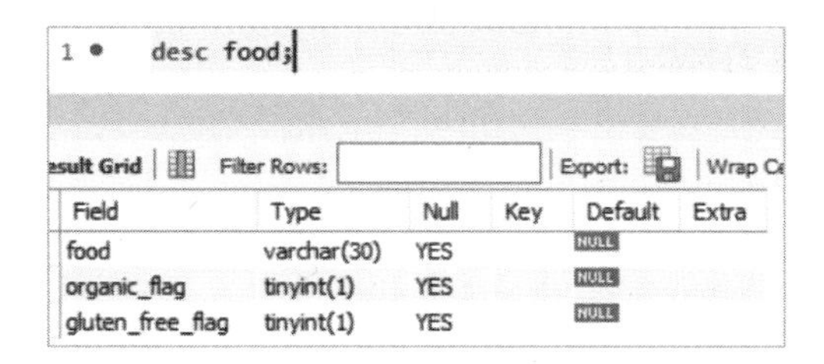

Figure 4-1: Describing the food *table in MySQL Workbench*

You can see that, although the `organic_flag` and `gluten_free_flag` columns were created with the `bool` synonym, the data type that was used to create those columns is `tinyint(1)`.

Decimal Data Types

For numbers that contain decimal points, MySQL provides the `decimal`, `float`, and `double` data types. Whereas `decimal` stores exact values, `float` and `double` store approximate values. For that reason, if you are storing values that can be handled equally well by `decimal`, `float`, or `double`, I recommend using the `decimal` data type.

`decimal`

The `decimal` data type allows you to define precision and scale. *Precision* is the total number of digits that you can store, and *scale* is the number of digits after the decimal point. The `decimal` data type is often used for monetary values with a scale of 2.

For example, if you define a `price` column as `decimal(5,2)`, you can store values between –999.99 and 999.99. A precision of `5` means you can store five total digits, and a scale of `2` means you can store two digits after the decimal point.

The following synonyms are available for the `decimal` type: `numeric(5,2)`, `dec(5,2)`, and `fixed(5,2)`. All of these are equivalent and create a data type of `decimal(5,2)`.

`float`

The `float` data type stores numeric data with a floating-point decimal. Unlike the `decimal` data type, where the scale is defined, a floating-point number has a decimal point that isn't always in the same location—the decimal point can *float* within the number. A `float` data type could represent the number 1.234, 12.34, or 123.4.

`double`

The `double` data type, short for `double precision`, also allows you to store a number with an undefined scale that has a decimal point someplace in the number. The `double` data type is similar to `float` except that `double` can store numbers more accurately. In MySQL, storing a `float` uses 4 bytes and storing a `double` uses 8. For floating-point numbers with many digits, use the `double` data type.

Date and Time Data Types

For dates and times, MySQL provides the `date`, `time`, `datetime`, `timestamp`, and `year` data types.

`date`

The `date` data type stores dates in `YYYY-MM-DD` format (year, month, and day, respectively).

`time`

The `time` data type stores times in `hh:mm:ss` format, representing hours, minutes, and seconds.

`datetime`

The `datetime` data type is for storing both the date and time in one value with the format `YYYY-MM-DD hh:mm:ss`.

`timestamp`

The `timestamp` data type also stores the date and the time in one value with the same format `YYYY-MM-DD hh:mm:ss`, though `timestamp` stores the *current* date and time, while `datetime` is designed for other date and time values.

The range of values that `timestamp` accepts is smaller; dates must be between the year 1970 and 2038. The `datetime` data type accepts a wider range of dates, from the years 1000 to 9999. You should use `timestamp` only when you want to stamp the current date and time value, such as to save the date and time that a row was updated.

`year`

The `year` data type stores the year in the `YYYY` format.

The json Data Type

JavaScript Object Notation (JSON) is a popular format for sending data between computers. MySQL provides the `json` data type to allow you to store and retrieve entire JSON documents in your database. MySQL will check that a JSON document contains valid JSON before allowing it to be saved in a `json` column.

A simple JSON document might look like this:

```
{
   "department":"Marketing",
   "city":"Detroit",
   "managers":[
      {
         "name":"Tom McBride",
         "age":29
      },
      {
         "name":"Jill Hatfield",
```

```
            "age":25
        }
    ]
}
```

JSON documents contain key/value pairs. In this example, `department` is a key and `Marketing` is a value. These keys and values don't correspond to rows and columns in your table; instead, the entire JSON document can be saved in a column that has the `json` data type. Later, you can extract properties from the JSON document using MySQL queries.

Spatial Data Types

MySQL provides data types for representing geographical location data, or *geodata*. This type of data helps answer questions like "What city am I in?" or "How many Chinese restaurants are within 5 miles of my location?"

`geometry`	Stores location values of any geographical type, including `point`, `linestring`, and `polygon` types
`point`	Represents a location with a particular latitude and longitude, like your current location
`linestring`	Represents points and the curve between them, such as the location of a highway
`polygon`	Represents a boundary, such as around a country or city
`multipoint`	Stores an unordered collection of `point` types
`multilinestring`	Stores a collection of `linestring` types
`emultipolygon`	Stores a collection of `polygon` typess
`geometrycollection`	Stores a collection of `geometry` types

TRY IT YOURSELF

4-1. Create a database named `rapper` and write a `create table` statement for the `album` table. The `album` table should have five columns:

- The `rapper_id` column should use an unsigned `smallint` data type.
- The `album_name` column should be a variable-length string that can hold up to 100 characters.
- The `explicit_lyrics_flag` should store a boolean value.
- The `album_revenue` column should store a monetary amount with a precision of 12 and a scale of 2.
- The `album_content` column should use the `longblob` data type.

Summary

In this chapter, you explored the available MySQL data types and when to use them. In the next chapter, you'll look at ways to retrieve data from multiple tables using different MySQL join types, and display that data in a single result set.

5

JOINING DATABASE TABLES

A SQL query walks into a bar, approaches two tables, and asks, "May I join you?"
—The worst database joke in history

Now that you've learned how to use SQL to select and filter data from a table, you'll see how to join database tables. *Joining* tables means selecting data from more than one table and combining it in a single result set. MySQL provides syntax to do different types of joins, like inner joins and outer joins. In this chapter, you'll look at how to use each type.

Selecting Data from Multiple Tables

The data you want to retrieve from a database often will be stored in more than one table, and you need to return it as one dataset in order to view all of it at once.

Let's look at an example. This table, called `subway_system`, contains data for every subway in the world:

```
subway_system              city              country_code
-------------------------  ----------------  ------------
Buenos Aires Underground   Buenos Aires      AR
Sydney Metro               Sydney            AU
Vienna U-Bahn              Vienna            AT
Montreal Metro             Montreal          CA
Shanghai Metro             Shanghai          CN
London Underground         London            GB
MBTA                       Boston            US
Chicago L                  Chicago           US
BART                       San Francisco     US
Washington Metro           Washington, D.C.  US
Caracas Metro              Caracas           VE
--snip--
```

The first two columns, `subway_system` and `city`, contain the name of the subway and the city where it's located. The third column, `country_code`, stores the two-character ISO country code. `AR` stands for Argentina, `CN` stands for China, and so on.

The second table, called `country`, has two columns, `country_code` and `country`:

```
country_code  country
------------  -----------
AR            Argentina
AT            Austria
AU            Australia
BD            Bangladesh
BE            Belgium
--snip--
```

Say you want to get a list of subway systems and their full city and country names. That data is spread across the two tables, so you'll need to join them to get the result set you want. Each table has the same `country_code` column, so you'll use that as a link to write a SQL query that joins the tables (see Listing 5-1).

```
select subway_system.subway_system,
       subway_system.city,
       country.country
from   subway_system
inner join country
on     subway_system.country_code = country.country_code;
```

Listing 5-1: Joining the subway_system and country tables

In the `country` table, the `country_code` column is the primary key. In the `subway_system` table, the `country_code` column is a foreign key. Recall that a primary key uniquely identifies rows in a table, and a foreign key is used to join with the primary key of another table. You use the `=` (equal) symbol to specify that you want to join all equal values from the subway_system and country tables' country_code columns.

Since you're selecting from two tables in this query, it's a good idea to specify which table the column is in every time you reference it, especially because the same column appears in both tables. There are two reasons for this. First, it will make the SQL easier to maintain because it will be immediately apparent in the SQL query which columns come from which tables. Second, because both tables have a column named country_code, if you don't specify the table name, MySQL won't know which column you want to use and will give an error message. To avoid this, in your select statement, type the table name, a period, and then the column name. For example, in Listing 5-1, subway_system.city refers to the city column in the subway_system table.

When you run this query, it returns all of the subway systems with the country names retrieved from the country table:

```
subway_system              city               country
------------------------   ----------------   -------------
Buenos Aires Underground   Buenos Aires       Argentina
Sydney Metro               Sydney             Australia
Vienna U-Bahn              Vienna             Austria
Montreal Metro             Montreal           Canada
Shanghai Metro             Shanghai           China
London Underground         London             United Kingdom
MBTA                       Boston             United States
Chicago L                  Chicago            United States
BART                       San Francisco      United States
Washington Metro           Washington, D.C.   United States
Caracas Metro              Caracas            Venezuela
--snip--
```

Note that the country_code column does not appear in the resulting join. This is because you selected only the subway_system, city, and country columns in the query.

When joining two tables based on columns with the same name, you can use the using *keyword instead of* on*. For example, replacing the last line in Listing 5-1 with* using (country_code); *would return the same result with less typing required.*

Table Aliasing

To save time when writing SQL, you can declare aliases for your table names. A *table alias* is a short, temporary name for a table. The following query returns the same result set as Listing 5-1:

```
select  s.subway_system,
        s.city,
        c.country
from    subway_system s
inner join country c
on      s.country_code = c.country_code;
```

You declare s as the alias for the subway_system table and c for the country table. Then you can type s or c instead of the full table name when referencing the column names elsewhere in the query. Keep in mind that table aliases are only in effect for the current query.

You can also use the word as to define table aliases:

```
select  s.subway_system,
        s.city,
        c.country
from    subway_system as s
inner join country as c
on      s.country_code = c.country_code;
```

The query returns the same results with or without as, but you'll cut down on typing by not using it.

Types of Joins

MySQL has several different types of joins, each of which has its own syntax, as summarized in Table 5-1.

Table 5-1: MySQL Join Types

Join type	Description	Syntax
Inner join	Returns rows where both tables have a matching value.	inner join join
Outer join	Returns all rows from one table and the matching rows from a second table. Left joins return all rows from the table on the left. Right joins return all rows from the table on the right.	left outer join left join right outer join right join
Natural join	Returns rows based on column names that are the same in both tables.	natural join
Cross join	Matches all rows in one table to all rows in another table and returns a Cartesian product.	cross join

Let's look at each type of join in more depth.

Inner Joins

Inner joins are the most commonly used type of join. In an inner join, there must be a match in both tables for data to be retrieved.

You performed an inner join on the subway_system and country tables in Listing 5-1. The returned list had no rows for Bangladesh and Belgium. These countries are not in the subway_system table, as they don't have subways; thus, there was not a match in both tables.

Note that when you specify inner join in a query, the word inner is optional because this is the default join type. The following query performs an inner join and produces the same results as Listing 5-1:

```
select  s.subway_system,
        s.city,
```

```
        c.country
from    subway_system s
join    country c
on      s.country_code = c.country_code;
```

You'll come across MySQL queries that use `inner join` and others that use `join`. If you have an existing codebase or written standards, it's best to follow the practices outlined there. If not, I recommend including the word `inner` for clarity.

Outer Joins

An outer join displays all rows from one table and any matching rows in a second table. In Listing 5-2, you select all countries and display subway systems for the countries if there are any.

```
select  c.country,
        s.city,
        s.subway_system
from    subway_system s right outer join country c
on      s.country_code = c.country_code;
```

Listing 5-2: Performing a right outer join

In this query, the `subway_system` table is considered the left table because it is to the left of the `outer join` syntax, while the `country` table is the right table. Because this is a *right* outer join, this query returns all the rows from the `country` table even if there is no match in the `subway_system` table. Therefore, all the countries appear in the result set, whether or not they have subway systems:

```
country               city           subway_system
--------------------  ------------   ------------------------
United Arab Emirates  Dubai          Dubai Metro
Afghanistan           null           null
Albania               null           null
Armenia               Yerevan        Yerevan Metro
Angola                null           null
Antarctica            null           null
Argentina             Buenos Aires   Buenos Aires Underground
--snip--
```

For countries without matching rows in the `subway_system` table, the `city` and `subway_system` columns display null values.

As with inner joins, the word `outer` is optional; using `left join` and `right join` will produce the same results as their longer equivalents.

The following outer join returns the same results as Listing 5-2, but uses the `left outer join` syntax instead:

```
select  c.country,
        s.city,
        s.subway_system
```

```
from    country c left outer join subway_system s
on      s.country_code = c.country_code;
```

In this query, the order of the tables is switched from Listing 5-2. The subway_system table is now listed last, making it the right table. The syntax country c left outer join subway_system s is equivalent to subway_system s right outer join country c in Listing 5-2. It doesn't matter which join you use as long as you list the tables in the correct order.

Natural Joins

A natural join in MySQL automatically joins tables when they have a column with the same name. Here is the syntax to automatically join two tables based on a column that is found in both:

```
select  *
from    subway_system s
natural join country c;
```

With natural joins, you avoid a lot of the extra syntax required for an inner join. In Listing 5-2, you had to include on s.country_code = c.country_code to join the tables based on their common country_code column, but with a natural join, you get that for free. The results of this query are as follows:

```
country_code   subway_system              city                country
------------   ------------------------   ------------        -------------
AR             Buenos Aires Underground   Buenos Aires        Argentina
AU             Sydney Metro               Sydney              Australia
AT             Vienna U-Bahn              Vienna              Austria
CA             Montreal Metro             Montreal            Canada
CN             Shanghai Metro             Shanghai            China
GB             London Underground         London              United Kingdom
US             MBTA                       Boston              United States
US             Chicago L                  Chicago             United States
US             BART                       San Francisco       United States
US             Washington Metro           Washington, D.C.    United States
VE             Caracas Metro              Caracas             Venezuela
--snip--
```

Notice that you selected all columns from the tables using the select * wildcard. Also, although both tables have a country_code column, MySQL's natural join was smart enough to display that column just once in the result set.

Cross Joins

MySQL's cross join syntax can be used to get the Cartesian product of two tables. A *Cartesian product* is a listing of every row in one table matched with every row in a second table. For example, say a restaurant has two database tables called main_dish and side_dish. Each table has three rows and one column.

The main_dish table is as follows:

```
main_item
---------
steak
chicken
ham
```

And the side_dish table looks like:

```
side_item
----------
french fries
rice
potato chips
```

A Cartesian product of these tables would be a list of all the possible combinations of main dishes and side dishes, and is retrieved using the cross join syntax:

```
select     m.main_item,
           s.side_item
from       main_dish m
cross join side_dish s;
```

This query, unlike the others you've seen, doesn't join tables based on columns. There are no primary keys or foreign keys being used. Here are the results of this query:

```
main_item   side_item
---------   ----------
ham         french fries
chicken     french fries
steak       french fries
ham         rice
chicken     rice
steak       rice
ham         potato chips
chicken     potato chips
steak       potato chips
```

Since there are three rows in the main_dish table and three rows in the side_dish table, the total number of possible combinations is nine.

Self Joins

Sometimes, it can be beneficial to join a table to itself, which is known as a self join. Rather than using special syntax as you did in the previous joins, you perform a self join by listing the same table name twice and using two different table aliases.

For example, the following table, called `music_preference`, lists music fans and their favorite genre of music:

```
music_fan   favorite_genre
---------   --------------
Bob         Reggae
Earl        Bluegrass
Ella        Jazz
Peter       Reggae
Benny       Jazz
Bunny       Reggae
Sierra      Bluegrass
Billie      Jazz
```

To pair music fans who like the same genre, you join the `music_preference` table to itself, as shown in Listing 5-3.

```
select a.music_fan,
       b.music_fan
from   music_preference a
inner join music_preference b
on (a.favorite_genre = b.favorite_genre)
where  a.music_fan != b.music_fan
order by a.music_fan;
```

Listing 5-3: Self join of the `music_preference` *table*

The `music_preference` table is listed twice in the query, aliased once as table `a` and once as table `b`. MySQL will then join tables `a` and `b` as if they are different tables.

In this query, you use the `!=` (not equal) syntax in the `where` clause to ensure that the value of the `music_fan` column from table `a` is not the same as the value of the `music_fan` column in table `b`. (Remember from Chapter 3 that you can use a `where` clause in your `select` statements to filter your results by applying certain conditions.) This way, music fans won't be paired up with themselves.

NOTE *The* `!=` *(not equal) syntax used here and the* `=` *(equal) syntax you've been using throughout this chapter are what's known as* comparison operators, *as they let you compare values in your MySQL queries. Chapter 7 will discuss comparison operators in more detail.*

Listing 5-3 produces the following result set:

```
music_fan  music_fan
---------  ---------
Benny      Ella
Benny      Billie
Billie     Ella
```

```
Billie    Benny
Bob       Peter
Bob       Bunny
Bunny     Bob
Bunny     Peter
Earl      Sierra
Ella      Benny
Ella      Billie
Peter     Bob
Peter     Bunny
Sierra    Earl
```

A music fan can now find other fans of their favorite genre in the right column next to their name.

NOTE *In Listing 5-3, the table is joined to itself as an inner join, but you could have used another type of join, like an outer join or a cross join.*

Variations on Join Syntax

MySQL allows you to write SQL queries that accomplish the same results in different ways. It's a good idea to get comfortable with different syntaxes, as you may have to modify code created by someone who doesn't write SQL queries in quite the same way that you do.

Parentheses

You can choose to use parentheses when joining on columns or leave them off. This query, which does not use parentheses

```
select  s.subway_system,
        s.city,
        c.country
from    subway_system as s
inner join country as c
on      s.country_code = c.country_code;
```

is the same as this query, which does:

```
select  s.subway_system,
        s.city,
        c.country
from    subway_system as s
inner join country as c
on      (s.country_code = c.country_code);
```

Both queries return the same result.

Old-School Inner Joins

This query, written in an older style of SQL, is equivalent to Listing 5-1:

```
select  s.subway_system,
        s.city,
        c.country
from    subway_system as s,
        country as c
where   s.country_code = c.country_code;
```

This code doesn't include the word join; instead, it lists the table names separated by a comma in the from statement.

When writing queries, use the newer syntax shown in Listing 5-1, but keep in mind that this older style is still supported by MySQL and you might see it used in some legacy code today.

Column Aliasing

You read earlier in the chapter about table aliasing; now you'll create aliases for columns.

In some parts of the world, like France, subway systems are referred to as *metros*. Let's select the subway systems for cities in France from the subway_system table and use column aliasing to display the heading metro instead:

```
select  s.subway_system as metro,
        s.city,
        c.country
from    subway_system as s
inner join country as c
on      s.country_code = c.country_code
where   c.country_code = 'FR';
```

As with table aliases, you can use the word as in your SQL query or you can leave it out. Either way, the results of the query are as follows, now with the subway_system column heading changed to metro:

```
metro           city        country
-----           --------    -------
Lille Metro     Lille       France
Lyon Metro      Lyon        France
Marseille Metro Marseille   France
Paris Metro     Paris       France
Rennes Metro    Rennes      France
Toulouse Metro  Toulouse    France
```

When creating tables, try to give your column headings descriptive names so that the results of your queries will be meaningful at a glance. In cases where the column names could be clearer, you can use a column alias.

Joining Tables in Different Databases

Sometimes there are tables with the same name in multiple databases, so you need to tell MySQL which database to use. There are a couple of different ways to do this.

In this query, the `use` command (introduced in Chapter 2) tells MySQL to use the specified database for the SQL statements that follow it:

```
use subway;

select * from subway_system;
```

On the first line, the `use` command sets the current database to `subway`. Then, when you select all the rows from the `subway_system` table on the next line, MySQL knows to pull data from the `subway_system` table in the `subway` database.

Here's a second way to specify the database name in your `select` statements:

```
select * from subway.subway_system;
```

In this syntax, the table name is preceded by the database name and a period. The `subway.subway_system` syntax tells MySQL that you want to select from the `subway_system` table in the `subway` database.

Both options produce the same result set:

```
subway_system        city                        country_code
-----------------    -------------------------   ------------
Buenos Aires         Underground Buenos Aires    AR
Sydney Metro         Sydney                      AU
Vienna U-Bahn        Vienna                      AT
Montreal Metro       Montreal                    CA
Shanghai Metro       Shanghai                    CN
London Underground   London                      GB
--snip--
```

Specifying the database and table name allows you to join tables that are in different databases on the same MySQL server, like so:

```
select  s.subway_system,
        s.city,
        c.country
from    subway.subway_system as s
inner join location.country as c
on      s.country_code = c.country_code;
```

This query joins the `country` table in the `location` database with the `subway_system` table in the `subway` database.

TRY IT YOURSELF

In the `solar_system` database, there are two tables: `planet` and `ring`. The `planet` table is as follows:

```
planet_id  planet_name
---------  ---------
    1      Mercury
    2      Venus
    3      Earth
    4      Mars
    5      Jupiter
    6      Saturn
    7      Uranus
    8      Neptune
```

The `ring` table stores only the planets with rings:

```
planet_id   ring_tot
---------   --------
    5           3
    6           7
    7          13
    8           6
```

5-1. Write a SQL query to perform an inner join between the `planet` and the `ring` tables, joining the tables based on their `planet_id` columns. How many rows do you expect the query to return?

5-2. Write a SQL query to do an outer join between the `planet` and the `ring` tables, with the `planet` table as the *left* table.

5-3. Modify your SQL query from Exercise 5-2 so that the `planet` table is the *right* table. The set returned by the query should be the same as the results of the previous exercise.

5-4. Modify your SQL query from Exercise 5-3 using a column alias. Make the `ring_tot` column display as `rings` in the heading of the result set.

Summary

In this chapter, you learned how to select data from two tables and display that data in a single result set using various joins offered by MySQL. In Chapter 6, you'll build on this knowledge by performing even more complex joins involving multiple tables.

6

PERFORMING COMPLEX JOINS WITH MULTIPLE TABLES

In Chapter 5, you saw how to join two tables and display the data in one result set. In this chapter, you'll create complex joins with more than two tables, learn about associative tables, and see how to combine or limit the results of a query. You'll then explore different ways to temporarily save a query's results in a table-like format, including temporary tables, derived tables, and Common Table Expressions (CTEs). Finally, you'll learn how to work with subqueries, which let you nest one query inside another for more refined results.

Writing One Query with Two Join Types

Joining three or more tables introduces greater complexity than joining two, as you might have different join types (like an inner and an outer join) in the same query. For example, Figure 6-1 illustrates three tables in the police database, which contains information on crimes, including the suspect and location.

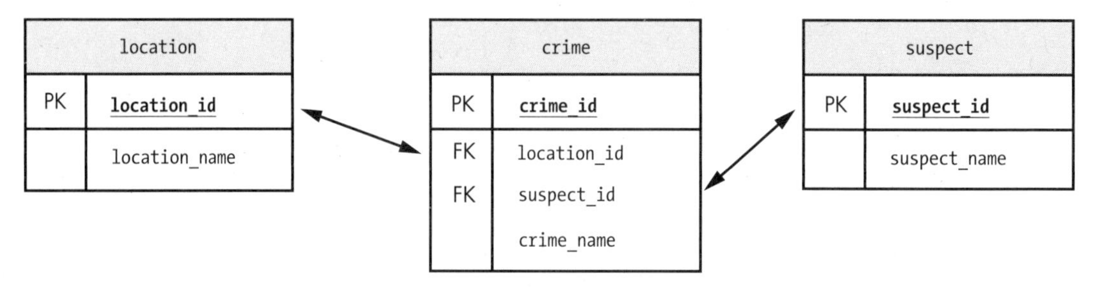

Figure 6-1: Three tables within the police *database*

The location table contains the locations where the crimes occurred:

```
location_id  location_name
-----------  -------------------------
    1        Corner of Main and Elm
    2        Family Donut Shop
    3        House of Vegan Restaurant
```

The crime table contains a description of the crimes:

```
crime_id  location_id  suspect_id  crime_name
--------  -----------  ----------  -------------------------------------
    1          1            1      Jaywalking
    2          2            2      Larceny: Donut
    3          3          null     Receiving Salad Under False Pretenses
```

The suspect table contains information about the suspect:

```
suspect_id  suspect_name
----------  ---------------
     1      Eileen Sideways
     2      Hugo Hefty
```

Say you want to write a query that joins all three tables to get a list of crimes, where they occurred, and the name of the suspect. The police database was designed so that there will always be a matching location in the location table for every crime in the crime table. However, there may not be a matching suspect in the suspect table because the police have not identified a suspect for every crime.

You'll perform an inner join between the crime and location tables, because you know there will be a match. But because there may not be a suspect match for each crime, you'll do an outer join between the crime table and the suspect table. Your query might look like this:

```
select c.crime_name,
       l.location_name,
       s.suspect_name
```

```
from   crime c
❶ join   location l
  on   c.location_id = l.location_id
❷ left join suspect s
  on   c.suspect_id = s.suspect_id;
```

In this example, you alias the tables with `c` for `crime`, `l` for `location`, and `s` for `suspect`. You use the `join` syntax for the inner join between the `crime` and `location` tables ❶, and the `left join` syntax for the outer join to the `suspect` table ❷.

Using a left join might cause some confusion in this context. When you were using `left join` with only two tables in Chapter 5, it was easy to understand which was the left table and which was the right, because there were only two possibilities. But how does it work now that you're joining three tables?

To understand multiple-table joins, imagine that MySQL is building temporary tables as it progresses through the query. MySQL joins the first two tables, `crime` and `location`, and the result of that join becomes the left table. Then MySQL does a `left join` between the `crime`/`location` combined table and the `suspect` table on the right.

You used a *left* join for the outer join because you want all of the crimes and locations to appear regardless of whether there is a match with the `suspect` table on the right. The results of this query are as follows:

```
crime_name                            location_name           suspect_name
------------------------------------  ----------------------  ---------------
Jaywalking                            Corner of Main and Elm  Eileen Sideways
Larceny: Donut                        Family Donut Shop       Hugo Hefty
Receiving Salad Under False Pretenses Green Vegan Restaurant  null
```

The suspect for the last crime was able to escape, so the value of the `suspect_name` on the last row is `null`. If you had used an inner join instead, the query wouldn't have returned the last row, because inner joins return rows only where there is a match.

You can use the null value that gets returned from an outer join to your advantage. Say you want to write a query to display only crimes where the suspect is not known. You could specify in the query that you want to see only rows where the suspect name is `null`:

```
select c.crime_name,
       l.location_name,
       s.suspect_name
from   crime c
join   location l
  on   c.location_id = l.location_id
left join suspect s
  on   c.suspect_id = s.suspect_id
where  s.suspect_name is null;
```

The results of this query are:

crime_name	location_name	suspect_name
Receiving Salad Under False Pretenses	Green Vegan Restaurant	null

Adding the where clause on the last line of the query showed you only rows that have no matching row in the suspect table, which limited your list to crimes with unknown suspects.

Joining Many Tables

MySQL allows up to 61 tables in a join, though you'll rarely need to write queries with that many. If you find yourself joining more than 10 tables, that's a sign the database could be redesigned to make writing queries simpler.

The wine database has six tables you can use to help plan a trip to a winery. Let's look at all six in turn.

The country table stores the countries where the wineries are located:

country_id	country_name
1	France
2	Spain
3	USA

The region table stores the regions within those countries where the wineries are located:

region_id	region_name	country_id
1	Napa Valley	3
2	Walla Walla Valley	3
3	Texas Hill	3

The viticultural_area table stores the wine-growing subregions where the wineries are located:

viticultural_area_id	viticultural_area_name	region_id
1	Atlas Peak	1
2	Calistoga	1
3	Wild Horse Valley	1

The wine_type table stores information about the types of wine available:

wine_type_id	wine_type_name
1	Chardonnay
2	Cabernet Sauvignon
3	Merlot

The winery table stores information about the wineries:

```
winery_id  winery_name           viticultural_area_id  offering_tours_flag
---------  --------------------  --------------------  -------------------
    1      Silva Vineyards                1                     0
    2      Chateau Traileur Parc          2                     1
    3      Winosaur Estate                3                     1
```

The portfolio table stores information about the winery's portfolio of wines—that is, which wines the winery offers:

```
winery_id  wine_type_id  in_season_flag
---------  ------------  --------------
    1            1             1
    1            2             1
    1            3             0
    2            1             1
    2            2             1
    2            3             1
    3            1             1
    3            2             1
    3            3             1
```

For example, the winery with a winery_id of 1 (Silva Vineyards) offers the wine with a wine_type_id of 1 (Chardonnay), which is in season (its in_season_flag—a boolean value—is 1, indicating true).

Listing 6-1 shows a query that joins all six tables to find a winery in the USA that has a Merlot in season and is offering tours.

```
select c.country_name,
       r.region_name,
       v.viticultural_area_name,
       w.winery_name
from   country c
join   region r
  on   c.country_id = r.country_id
 and   c.country_name = 'USA'
join   viticultural_area v
  on   r.region_id = v.region_id
join   winery w
  on   v.viticultural_area_id = w.viticultural_area_id
 and   w.offering_tours_flag is true
join   portfolio p
  on   w.winery_id = p.winery_id
 and   p.in_season_flag is true
join   wine_type t
  on   p.wine_type_id = t.wine_type_id
 and   t.wine_type_name = 'Merlot';
```

Listing 6-1: A query to list US wineries with in-season Merlot

While this is a longer query than you're used to, you've seen most of the syntax before. You create table aliases for each table name in the query (country, region, viticultural_area, winery, portfolio, and wine_type). When referring to columns in the query, you precede the column names with the table aliases and a period. For example, you precede the offering_tours_flag column with w because it is in the winery table, resulting in w.offering_tours_flag. (Remember from Chapter 4 that it's best practice to add the suffix _flag to columns that contain boolean values like true or false, which is the case with the offering_tours column, since a winery either offers tours or doesn't.) Finally, you perform inner joins on each table with the word join, as there should be matching values when you join these tables.

Unlike our earlier queries, this query contains some joins between tables where more than one condition must be met. For example, when you join the country and region tables, there are *two* conditions that need to be met:

- The value in the country_id column of the country table must match the value in the country_id column of the region table.
- The value in the country_name column of the country table must equal USA.

You handled the first condition using the on keyword:

```
from   country c
join   region r
  on   c.country_id = r.country_id
```

Then you used the and keyword to specify the second condition:

```
 and   c.country_name = 'USA'
```

You can add more and statements to specify as many joining conditions as you need.

The results of the query in Listing 6-1 are as follows:

```
country_name   region_name     viticultural_area_name   winery_name
------------   -------------   ----------------------   --------------------
    USA        Napa Valley     Calistoga                Chateau Traileur Parc
    USA        Napa Valley     Wild Horse Valley        Winosaur Estate
```

Associative Tables

In Listing 6-1, most of the tables are straightforward: the winery table stores a list of wineries, region stores a list of regions, country stores countries, and viticultural_area stores viticultural areas (wine-growing subregions).

The portfolio table, however, is a little different. Remember, it stores information about which wines are in each winery's portfolio. Here it is again:

```
winery_id  wine_type_id  in_season_flag
---------  ------------  --------------
    1            1              1
    1            2              1
```

```
1             3             0
2             1             1
2             2             1
2             3             1
3             1             1
3             2             1
3             3             1
```

Its winery_id column is the primary key of the winery table, and its wine_type_id column is the primary key of the wine_type table. This makes portfolio an *associative table* because it associates rows that are stored in other tables to each other by referencing their primary keys, as illustrated in Figure 6-2.

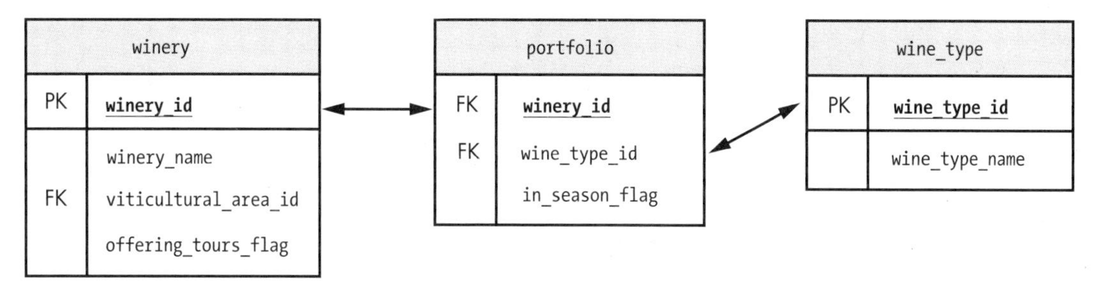

Figure 6-2: The portfolio *table is an associative table.*

The portfolio table represents *many-to-many relationships* because one winery can produce many wine types, and one wine type can be produced in many wineries. For example, winery 1 (Silva Vineyards) offers many wine types: 1 (Chardonnay), 2 (Cabernet Sauvignon), and 3 (Merlot). Wine type 1 (Chardonnay) is offered by many wineries: 1 (Silva Vineyards), 2 (Chateau Traileur Parc), and 3 (Winosaur Estate). The portfolio table contains that list of relationships between each winery_id and wine_type_id that tells us which wineries have which wine types. As a bonus, it also contains the in_season_flag column, which, as you've seen, tracks whether that wine is in season at that winery.

Next, we'll look at different ways to work with the data that's returned from your queries. We'll start with some simple options for managing the data in your result set and then cover some more involved approaches in the latter half of the chapter.

Managing the Data in Your Result Set

Sometimes you'll want to control how much data from your queries is displayed in your result set. For example, you might want to pare down your results or combine the results of several select statements. SQL provides keywords to add this functionality to your queries.

The limit Keyword

The `limit` keyword lets you limit the number of rows displayed in your result set. For example, consider a table called `best_wine_contest` that holds the results of a contest where wine tasters voted for their favorite wines. If you query the table and `order by` the `place` column, you'll see the wines that ranked the best first:

```
select *
from   best_wine_contest
order by place;
```

The results are:

```
wine_name      place
------------   -----
Riesling         1
Pinot Grigio     2
Zinfandel        3
Malbec           4
Verdejo          5
```

If you want to see only the top three wines, use `limit 3`:

```
select *
from   best_wine_contest
order by place
limit 3;
```

Now the results are:

```
wine_name      place
------------   -----
Riesling         1
Pinot Grigio     2
Zinfandel        3
```

The `limit` keyword limited the results to three rows. To see only the wine that won top place, you could use `limit 1`.

The union Keyword

The `union` keyword combines the results of multiple `select` statements into one result set. For example, the following query selects all the wine types from two different tables, `wine_type` and `best_wine_contest`, and shows them in one list:

```
select wine_type_name from wine_type
union
select wine_name from best_wine_contest;
```

The result is:

```
wine_type_name
------------------
Chardonnay
Cabernet Sauvignon
Merlot
Riesling
Pinot Grigio
Zinfandel
Malbec
Verdejo
```

The `wine_type` table has a column called `wine_type_name` that includes Chardonnay, Cabernet Sauvignon, and Merlot. The `best_wine_contest` table has a column called `wine_name` that includes Riesling, Pinot Grigio, Zinfandel, Malbec, and Verdejo. Using `union` allows you to see all of the wines together in one result set.

You can use `union` only when every `select` statement has the same number of columns. The union works in this example because you specified just one column in each of the `select` statements. The column name in the result set is usually taken from the first `select` statement.

The `union` keyword will remove duplicate values from the result set. For example, if you had Merlot in both the `wine_type` and the `best_wine_contest` tables, using `union` would produce a list of distinct wines, with Merlot listed only once. To see a list that includes duplicate values, use `union all`:

```
select wine_type_name from wine_type
union all
select wine_name from best_wine_contest;
```

The result would be:

```
wine_type_name
------------------
Chardonnay
Cabernet Sauvignon
Merlot
Riesling
Pinot Grigio
Zinfandel
Malbec
Verdejo
Merlot
```

Now you can see that Merlot is listed twice.

Next, you'll dive in a bit deeper to make your queries even more efficient by creating temporary result sets in a table-like format.

TRY IT YOURSELF

In the nutrition database, there are two tables, good_snack and bad_snack.

The good_snack table looks like this:

```
snack_name
----------
carrots
salad
soup
```

The bad_snack table looks like this:

```
snack_name
----------
sausage pizza
BBQ ribs
nachos
```

6-1. Write a query to display all the snacks from both tables in one result set.

Temporary Tables

MySQL allows you to create temporary tables—that is, a temporary result set that will exist only for your current session and then be automatically dropped. For example, you can create a temporary table using a tool like MySQL Workbench and then query that table within the tool. If you close and reopen MySQL Workbench, however, the temporary table will be gone. You can reuse a temporary table several times in a single session.

You can define a temporary table the same way you create a regular table, except that you use the syntax `create temporary table` instead of `create table`:

```
create temporary table wp1
(
    winery_name              varchar(100),
    viticultural_area_id     int
)
```

The `wp1` temporary table gets created with the column names and data types that you specified, without any rows.

To create a temporary table based on the results of a query, simply precede the query with the same `create temporary table` syntax, as shown in Listing 6-2, and the resulting temporary table will contain the rows of data that were selected from the query.

```
create temporary table winery_portfolio
select w.winery_name,
       w.viticultural_area_id
from   winery w
join   portfolio p
❶ on   w.winery_id = p.winery_id
❷ and  w.offering_tours_flag is true
  and  p.in_season_flag is true
join   wine_type t
❸ on   p.wine_type_id = t.wine_type_id
❹ and  t.wine_type_name = 'Merlot';
```

Listing 6-2: Creating a temporary table

Here you create a temporary table called `winery_portfolio` that stores the results of a query joining the `winery`, `portfolio`, and `wine_type` tables from Listing 6-1 and Figure 6-2. The `winery` and `portfolio` tables are joined based on two conditions:

- The values of the `winery_id` columns in the tables match ❶.
- The winery is offering tours. For this, you check that the `offering_tours_flag` in the `winery` table is set to `true` ❷.

Those results are joined with the `wine_type` table based on two conditions:

- The values of the `wine_type_id` columns in the tables match ❸.
- The `wine_type_name` in the `wine_type` table is `Merlot` ❹.

NOTE *Temporary tables are created with the data types of the columns you've selected in the query. For example, in Listing 6-2 you selected `winery_name` from the `winery` table, which was defined as `varchar(100)`, so the `winery_portfolio` temporary table also gets created with a `winery_name` column defined as `varchar(100)`.*

Once you've created a temporary table, you can query its contents by selecting from it, just as you would with a permanent table:

```
select * from winery_portfolio;
```

The results are:

```
winery_name            viticultural_area_id
---------------------  --------------------
Chateau Traileur Parc           2
Winosaur Estate                 3
```

Now you can write a second query to select from the `winery_portfolio` temporary table and join it with other three tables from Listing 6-1:

```
select c.country_name,
       r.region_name,
       v.viticultural_area_name,
```

```
       w.winery_name
from   country c
join   region r
  on   c.country_id = r.country_id
 and   c.country_name = 'USA'
join   viticultural_area v
  on   r.region_id = v.region_id
join   winery_portfolio w
  on   v.viticultural_area_id = w.viticultural_area_id;
```

Here you are joining the `winery_portfolio` temporary table to the remaining tables that were part of the original query in Listing 6-1: `country`, `region`, and `viticultural_area`. In this way, you simplified a large, six-table query by isolating the data from three tables into a temporary table and then joining that temporary table with the other three tables. This query returns the same results as Listing 6-1.

TRY IT YOURSELF

The `canada` database contains the `province`, `capital_city`, and `tourist_attraction` tables shown here.

The `province` table looks like this:

```
province_id  province_name          official_language
-----------  ---------------------  -----------------
     1       Alberta                English
     2       British Columbia       English
     3       Manitoba               English
     4       New Brunswick          English, French
     5       Newfoundland           English
     6       Nova Scotia            English
     7       Ontario                English
     8       Prince Edward Island   English
     9       Quebec                 French
    10       Saskatchewan           English
```

The `capital_city` table looks like this:

```
city_id  city_name     province_id
-------  ---------     -----------
   1     Toronto            7
   2     Quebec City        9
   3     Halifax            5
   4     Fredericton        4
   5     Winnipeg           3
   6     Victoria           2
   7     Charlottetown      8
   8     Regina            10
   9     Edmonton           1
  10     St. Johns          5
```

The tourist_attraction table looks like this:

attraction_id	attraction_name	attraction_city_id	open_flag
1	CN Tower	1	true
2	Old Quebec	2	true
3	Royal Ontario Museum	1	true
4	Place Royale	2	true
5	Halifax Citadel	3	true
6	Garrison District	4	true
7	Confederation Centre of...	7	true
8	Stone Hall Castle	8	true
9	West Edmonton Mall	9	true
10	Signal Hill	10	true
11	Canadian Museum for Human...	5	true
12	Royal BC Museum	6	true
13	Sunnyside Amusement Park	1	false

6-2. Write a query that performs an inner join between the three tables. Select the attraction_name column from the tourist_attraction table, the city_name column from the capital_city table, and the province_name column from the province table.

Select rows from the attraction table only where the open_flag is set to true. Select rows from the province table only where the official_language is set to French.

6-3. Create a temporary table called open_tourist_attraction that selects the attraction_city_id and attraction_name columns from the tourist_attraction table where the open_flag value is true.

6-4. Write a query that joins the open_tourist_attraction temporary table you created in Exercise 6-2 to the capital_city table. Select the attraction_name column from the open_tourist_attraction temporary table and the city_name column from the capital_city table. Select only rows from the capital_city table that have a city_name of Toronto.

Common Table Expressions

Common Table Expressions (CTEs), a feature introduced in MySQL version 8.0, are a temporary result set that you name and can then select from as if it were a table. You can use CTEs only for the duration of one query (versus temporary tables, which can be used for the entire session). Listing 6-3 shows how to use a CTE to simplify the query from Listing 6-1:

```
❶ with winery_portfolio_cte as
  (
     select w.winery_name,
            w.viticultural_area_id
     from   winery w
```

```
      join   portfolio p
        on   w.winery_id = p.winery_id
       and   w.offering_tours_flag is true
       and   p.in_season_flag is true
      join   wine_type t
        on   p.wine_type_id = t.wine_type_id
       and   t.wine_type_name = 'Merlot'
)
❷ select c.country_name,
         r.region_name,
         v.viticultural_area_name,
         wp.winery_name
  from   country c
  join   region r
    on   c.country_id = r.country_id
   and   c.country_name = 'USA'
  join   viticultural_area v
    on   r.region_id = v.region_id
❸ join   winery_portfolio_cte wp
    on   v.viticultural_area_id = wp.viticultural_area_id;
```

Listing 6-3: Naming and then querying a CTE

First, you use the `with` keyword to give the CTE a name; here, you define the name `winery_portfolio_cte` for the results of the query shown between the parentheses ❶. Then you add another query ❷ that uses `winery_portfolio_cte` in a join as if it were a table ❸. The results are the same as those of Listing 6-1.

CTEs and temporary tables both temporarily save the results of a query in a table-like format. However, while temporary tables can be used more than once (that is, in multiple queries) in a session, CTEs can be used only for the duration of the query in which they are defined. After you run Listing 6-3, try to run another query to select from `winery_portfolio_cte`:

```
select * from winery_portfolio_cte;
```

You'll get an error:

```
Error Code: 1146. Table 'wine.winery_portfolio_cte' doesn't exist
```

MySQL is looking for a *table* named `winery_portfolio_cte`, so it's no surprise it can't locate your CTE. Besides that, the CTE existed only for the duration of your query, so it's no longer available.

Recursive Common Table Expressions

Recursion is a technique that is used when an object references itself. When I think of recursion, I think of Russian nesting dolls. You open the largest doll and discover a smaller doll within it; then you open that doll and find an even smaller doll within that; and so on until you reach the tiniest doll in the center. In other words, to see all the dolls, you start with the largest doll and then iterate through each smaller doll until you find a doll that doesn't contain another one.

Recursion is useful when your data is organized as a hierarchy or a series of values where you need to know the previous value to arrive at the current value.

A recursive CTE references itself. Recursive CTEs have two `select` statements separated by a `union` statement. Take a look at this recursive CTE called `borg_scale_cte`, which contains a series of numbers between 6 and 20:

```
❶ with recursive borg_scale_cte as
(
  ❷ select    6 as current_count
    union
  ❸ select    current_count + 1
    from      borg_scale_cte
  ❹ where     current_count < 20
)
select * from borg_scale_cte;
```

First, you define the CTE as `recursive` and name it `borg_scale_cte` ❶. Then, the first `select` statement returns the first row containing the number `6` ❷. The second `select` statement returns all other rows with values `7` through `20`. It continually adds `1` to the `current_count` column and selects the resulting numbers ❸, so long as the `current_count` is less than `20` ❹.

In the last line, you use the wildcard character * to select all the values from the CTE, which returns:

```
current_count
-------------
      6
      7
      8
      9
     10
     11
     12
     13
     14
     15
     16
     17
     18
     19
     20
```

You can also use a recursive CTE as if it were a table and join it with other tables, for example.

Derived Tables

Derived tables are an alternative to CTEs for creating a table of results just for use within a query. The SQL that creates the derived table goes within parentheses:

```
select   wot.winery_name,
         t.wine_type_name
from     portfolio p
join wine_type t
on       p.wine_type_id = t.wine_type_id
join (
     select *
     from   winery
     where  offering_tours_flag is true
     ) wot
On       p.winery_id = wot.winery_id;
```

The query within the parentheses produces a derived table aliased as `wot` (short for *wineries offering tours*). You can treat `wot` as if it were just another table, joining it to the `portfolio` and `wine_type` tables, and selecting columns from it. As with a CTE, the derived table is available just for the duration of your query.

The choice to use a derived table rather than a CTE is often a matter of style. Some developers prefer to use CTEs because they feel CTEs are a more readable option. If you need to use recursion, however, you would have to use a CTE.

Subqueries

A subquery (or inner query) is a query nested within another query. A subquery is used to return data that will be used by the main query. When a query has a subquery, MySQL runs the subquery first, selects the resulting value from the database, and then passes it back to the outer query. For example, this SQL statement uses a subquery to return a list of all the wine-growing regions in the United States from the `wine` database:

```
❶ select region_name
  from   region
  where  country_id =
  (
   ❷ select country_id
     from   country
     where  country_name = 'USA'
  );
```

The result of this query is as follows:

```
region_name
------------------
Napa Valley
Walla Walla Valley
Texas Hill
```

The query has two parts: the outer query ❶ and the subquery ❷. Try running the subquery in isolation, without the outer query:

```
    select country_id
    from   country
    where  country_name = 'USA';
```

The result shows that the `country_id` returned for USA is `3`:

```
country_id
----------
     3
```

In your query, `3` is passed from the subquery to the outer query, which makes the entire SQL statement evaluate to:

```
select region_name
from   region
where  country_id = 3;
```

This results in a list of regions for `country_id 3 (USA)` being returned:

```
region_name
------------------
Napa Valley
Walla Walla Valley
Texas Hill
```

Subqueries That Return More Than One Row

Subqueries can return more than one row. Here's the same query as before, this time including all countries, not just the USA:

```
  select region_name
  from   region
❶ where  country_id =
  (
      select country_id
      from   country
❷ --  where  country_name = 'USA' -  line commented out
  );
```

Now the line of the subquery that specifies that you want only USA regions is commented out ❷, so the country_id for all countries will be returned. When you run this query, instead of a list of regions, MySQL returns an error:

```
Error Code: 1242. Subquery returns more than 1 row
```

The problem is that the outer query expects only one row to be returned because you used the = syntax ❶. Instead, the subquery returns three rows: country_id 3 for the USA, 1 for France, and 2 for Spain. You should use = only when there is no possibility that the subquery could return more than one row.

This is a common mistake that you should be mindful of. Many developers have written a query that worked when they tested it, but all of a sudden, one day it starts producing Subquery returns more than 1 row errors. Nothing has changed about the query (unlike in this case where a line has been commented out), but the data in their database has changed. For example, new rows might have been added to a table, and the developer's subquery now returns multiple rows where it used to return one.

To write a query where more than one row can be returned from the subquery, you can use the in keyword instead of =:

```
select region_name
from   region
where  country_id in
(
    select country_id
    from   country
--  where  country_name = 'USA' -  line commented out
);
```

Now that you've replaced = with in, the outer query can accept multiple rows back from the subquery without error, and you'll get a list of regions for all countries.

TRY IT YOURSELF

Say you're working at a staffing firm and you receive a report about a query that used to run successfully but is now failing with an error message of Subquery returns more than 1 row. You've been asked to fix this query:

```
select  employee_id,
        hat_size
from    wardrobe
where   employee_id =
(
        select   employee_id
        from     employee
```

```
        where    position_name = 'Pope'
);
```

The attire database contains the wardrobe and employee tables. The wardrobe table looks like this:

```
employee_id  hat_size
-----------  --------
      1        8.25
      2        7.50
      3        6.75
```

The employee table looks like this:

```
employee_id  employee_name  position_name
-----------  -------------  -------------
     1       Benedict       Pope
     2       Garth          Singer
     3       Francis        Pope
```

6-5. Employee 3 has been added to the database recently. How would you fix the query? Why do you think the query used to work without a problem, but now the same query fails?

Correlated Subqueries

In a correlated subquery, a column from a table in the subquery is joined with a column from a table in the outer query.

Let's take a look at two tables called best_paid and employee in the pay database. The best_paid table shows that the highest salary in the Sales department is $200,000, and the highest salary in the Manufacturing department is $80,000:

```
department     salary
----------     ------
Sales          200000
Manufacturing   80000
```

The employee table stores a list of employees, their department, and their salary:

```
employee_name  department     salary
-------------  -------------  ------
Wanda Wealthy  Sales          200000
Paul Poor      Sales           12000
Mike Mediocre  Sales           70000
Betty Builder  Manufacturing   80000
Sean Soldering Manufacturing   80000
Ann Assembly   Manufacturing   65000
```

You can use a correlated subquery to find the highest-paid employees in each department:

```
select employee_name,
       salary
from   employee e
where  salary =
       (
       select  b.salary
       from    best_paid b
       where   b.department = e.department
       );
```

In the outer query, you select employees and salaries from the `employee` table. In the subquery, you join the results of the outer query with the `best_paid` table to determine if this employee has the highest salary for their department.

The results are:

```
employee_name   salary
-------------   ------
Wanda Wealthy   200000
Betty Builder    80000
Sean Soldering   80000
```

The results show that Wanda is the highest-paid employee in the Sales department and Betty and Sean are tied for the highest salary in the Manufacturing department.

TRY IT YOURSELF

In the `monarchy` database, there is a table named `royal_family` that contains the following data:

```
name                               birthdate
--------------------------------   ----------
Prince Louis of Cambridge          2018-04-23
Princess Charlotte of Cambridge    2015-05-02
Prince George of Cambridge         2013-07-22
Prince William, Duke of Cambridge  1982-06-21
Catherine, Duchess of Cambridge    1982-01-09
Charles, Prince of Whales          1948-11-14
Queen Elizabeth II                 1926-04-21
Prince Andrew, Duke of York        1960-02-19
```

6-6. Write a query to select all columns from the table and order by the `birthdate` column.

6-7. Add `limit 1` to the end of the query to see who is the oldest royal in the table.

6-8. Now change the query to order by the `birthdate` column in descending order. Then add `limit 3` to the end of the query to see who the youngest three royals in the table are.

Summary

In this chapter, you wrote complex SQL statements using multiple tables. You saw how to limit or combine the rows of your results, and you explored several different ways to write queries using result sets as if they were tables.

In the next chapter, you'll compare values in your queries; for example, you'll check that one value is more than another value, compare values that have different data types, and check whether a value matches some pattern.

7

COMPARING VALUES

This chapter discusses comparing values in MySQL. You'll practice checking whether values are equal, whether one value is greater or less than another value, and whether a value falls within a specific range or matches a pattern. You'll also learn how to check that at least one condition in your queries is met.

Comparing values can be useful in a variety of scenarios. For example, you might want to check that an employee worked 40 or more hours, that a flight's status is not canceled, or that the average temperature of a vacation destination is between 70 and 95 degrees Fahrenheit.

Comparison Operators

You can use MySQL's comparison operators, shown in Table 7-1, to compare values in your queries.

Table 7-1: MySQL Comparison Operators

Symbol or keyword(s)	Description
=	Equal
!=, <>	Not equal
>	Greater than
>=	Greater than or equal to
<	Less than
<=	Less than or equal to
is null	A null value
is not null	A non-null value
in	Matches a value in a list
not in	Doesn't match a value in a list
between	Within a range
not between	Not within a range
like	Matches a pattern
not like	Does not match a pattern

These operators let you compare values in a database to other values. You can choose to select data if it meets the criteria you define using these comparison operators. Let's discuss them in depth, using various databases as examples.

Equal

The equal operator, introduced in Chapter 5, lets you check that values are equal to each other to achieve specific results. For example, here you use = with the wine database table from Chapter 6:

```
select  *
from    country
where   country_id = 3;
```

This query selects all countries from the country table that have a country_id equal to 3.

In the following query, you're using = with a string, rather than a number:

```
select  *
from    wine_type
where   wine_type_name = 'Merlot';
```

This query selects all wines from the wine_type table with the name Merlot—that is, a wine_type_name equal to Merlot.

The following query is similar to what you saw in Chapter 5 when you were learning how to join two tables. Here you're using = to compare values that come from two tables with a common column name:

```
select  c.country_name
from    country c
join    region r
  on    c.country_id = r.country_id;
```

This query joins all equal values from the `region` and `country` tables' `country_id` columns.

In each of these examples, the = syntax checks that the value on the left of the operator is the same as the value on the right of it. You can also use = with a subquery that returns one row:

```
select *
from   region
where  country_id =
(
    select country_id
    from   country
    where  country_name = 'USA'
);
```

Using = in this way, you're checking for rows in the outer query where the `country_id` column in the `region` table matches the results of an entire subquery.

NOTE *Regardless of the comparison operator that you use, you should compare values that have the same data type. For example, you should avoid comparing an* `int` *with a* `varchar` *in your queries. In some cases, MySQL can perform automatic conversions, but this is not best practice.*

Not Equal

Not equal is expressed by the <> or != symbols, where the < symbol is *less than* and the > symbol is *greater than* (so <> means less than or greater than), and the ! symbol means *not* (so != means not equal). The != and <> operators do the same thing, so it doesn't matter which syntax you use.

The not equal operator is useful for excluding certain data from the results of your queries. For example, maybe you're a banjo player looking for fellow musicians to start a band. Since you play banjo, you can eliminate it from the list of instruments you want to see:

```
select  *
from    musical_instrument
where   instrument != 'banjo';
```

Here you've used the `not equal` operator on the `musical_instrument` table to exclude the banjo from the list of instruments returned.

Say you're planning a wedding and you have a prior commitment on February 11, 2024, so you need to exclude that date:

```
select   *
from     possible_wedding_date
where    wedding_date <> '2024-02-11';
```

Now you've excluded 2/11/2024 from a list of potential wedding dates in your `possible_wedding_date` table.

Greater Than

The greater than operator checks that the value on the left is greater than the value on the right. It is expressed using the `>` symbol. Say you're looking for jobs that have a `salary` greater than $100,000 and a `start_date` after 1/20/2024. You can select jobs that match these requirements from the `job` table using the following query:

```
select   *
from     job
where    salary > 100000
and      start_date > '2024-01-20';
```

In this query, only the jobs that meet both conditions will be returned.

Greater Than or Equal To

Greater than or equal to is expressed using the `>=` symbol. For example, you can edit your previous query to select all jobs where the `salary` is $100,000 or higher and that have a `start_date` of 1/20/2024 or later:

```
select   *
from     job
where    salary >= 100000
and      start_date >= '2024-01-20';
```

The difference between `>` and `>=` is that `>=` includes the value listed in its results. In the previous examples, a job with a `salary` of *exactly* $100,000 will be returned by `>=` but not by `>`.

Less Than

Less than is expressed using the `<` symbol. For example, to view all games starting before 10 PM, you can perform the following query:

```
select *
from   team_schedule
where  game_time < '22:00';
```

In MySQL, time is expressed in military format, which operates on a 24-hour clock.

Less Than or Equal To

Less than or equal to is expressed using the <= symbol. You can expand the previous query to select all rows where the game_time is 10 PM or earlier:

```
select *
from   team_schedule
where  game_time <= '22:00';
```

If the game_time is exactly 22:00 (10 PM), a row will be returned when you use <= but not when you use <.

is null

As discussed in Chapters 2 and 3, null is a special value indicating that data is not applicable or not available. The is null syntax allows you to specify that you want only null values to be returned from a table. For example, say you want to query the employee table to see a list of employees who have not retired or set a retirement date:

```
select  *
from    employee
where   retirement_date is null;
```

Now only rows with a retirement_date of null are returned:

```
emp_name   retirement_date
--------   ---------------
Nancy      null
Chuck      null
Mitch      null
```

It's only possible to check null values with the is null comparison operator. For example, using = null won't work:

```
select *
from   employee
where  retirement_date = null;
```

Even though there are null values in the table, this syntax won't return any rows. In this scenario, MySQL doesn't throw an error, so you might not realize that the wrong data is being returned.

is not null

You can use is not null to check for values that are *not* null. Try reversing the logic of the previous example to check for employees who have retired or set a retirement date:

```
select *
from   employee
where  retirement_date is not null;
```

Now, the query returns rows with a retirement_date that is not null:

```
emp_name   retirement_date
--------   ---------------
Alfred     2034-01-08
Latasha    2029-11-17
```

As with is null, you have to use the is not null syntax for this type of query. Using other syntax, like != null or <> null, will not produce the correct results:

```
select *
from   employee
where  retirement_date != null;
```

As you saw earlier with = null, MySQL won't return any rows when you try to use the != null syntax, and won't alert you with an error.

In

You can use the in keyword to specify a list of multiple values you want your query to return. For example, let's revisit the wine database to return specific wines from the wine_type table:

```
select   *
from     wine_type
where    wine_type_name in ('Chardonnay', 'Riesling');
```

This will return rows where the wine_type_name is Chardonnay or Riesling.

You can also use in with a subquery to select a list of wine types that are in another table:

```
select   *
from     wine_type
where    wine_type_name in
         (
         select  wine_type_name
         from    cheap_wine
         );
```

Instead of providing a hardcoded list of wine types to return in your results, here you're selecting all of the wine types from the cheap_wine table.

not in

To reverse the previous example's logic and exclude certain wine types, you can use not in:

```
select   *
from     wine_type
where    wine_type_name not in ('Chardonnay', 'Riesling');
```

This returns all rows where the `wine_type_name` is not Chardonnay or Riesling.

To select wines that are not from the `cheap_wine` table, you can use `not in` within a subquery as follows:

```
select  *
from    wine_type
where   wine_type_name not in
        (
        select  wine_type_name
        from    cheap_wine
        );
```

This query excludes wine types from the `cheap_wine` table.

between

You can use the `between` operator to check that a value is within a specified range. For example, to list the millennials in a `customer` table, search for people who were born between 1981 and 1996:

```
select  *
from    customer
where   birthyear between 1981 and 1996;
```

The `between` keyword is *inclusive*. This means it checks for every `birthyear` within the range, *including* the years 1981 and 1996.

not between

You can check that a value is not within a range by using the `not between` operator. Use the same table from the previous example to find customers who are *not* millennials:

```
select  *
from    customer
where   birthyear not between 1981 and 1996;
```

The `not between` operator returns the opposite list of customers that `between` did, and is *exclusive*. Customers born in 1981 or 1996 will be *excluded* by this query since they are part of the `between 1981 and 1996` group.

like

The `like` operator allows you to check if a string matches some pattern. For example, you can use `like` to find books from No Starch Press by checking if a book's ISBN contains the No Starch publisher code, 59327.

To specify the pattern to match, you use one of two wildcard characters with the `like` operator: percent (`%`) or underscore (`_`).

The % Character

The percent wildcard character matches any number of characters. For example, to return a list of billionaires whose last name starts with the letter *M*, you can use the % wildcard character along with like:

```
select  *
from    billionaire
where   last_name like 'M%';
```

Your query will find billionaires whose last name starts with an *M* followed by zero or more other characters. This means that like 'M%' would match only the letter *M* with no characters after it, or *M* followed by a few characters, like Musk, or *M* followed by many characters, like Melnichenko. The results of your query might look like this:

```
first_name  last_name
----------  ---------
Elon        Musk
Jacqueline  Mars
John        Mars
Andrey      Melnichenko
```

You can use two % characters to find a character located anywhere in the string, whether at the beginning, in the middle, or at the end. For example, the following query looks for billionaires whose last names contain the letter *e*:

```
select  *
from    billionaire
where   last_name like '%e%';
```

The results might look like this:

```
first_name  last_name
----------  ---------
Jeff        Bezos
Bill        Gates
Mark        Zuckerberg
Andrey      Melnichenko
```

While the syntax last_name like '%e%' is handy, it can cause your query to run slower than normal. That's because when you use the % wildcard at the beginning of a search pattern, MySQL can't take advantage of any indexes on the last_name column. (Remember, indexes help MySQL optimize your queries; for a refresher, see the section "Indexes" in Chapter 2.)

The _ Character

The underscore wildcard character matches any character. For example, say you need to find a contact and you can't remember if her name was Jan or Jen. You might write a query to select names that start with *J*, followed by the wildcard character, followed by *n*.

Here you use the underscore wildcard to return a list of three-letter terms that end in *at*:

```
select  *
from    three_letter_term
where   term like '_at';
```

The results might look like this:

```
term
----
cat
hat
bat
```

> **TRY IT YOURSELF**
>
> **7-1.** In the band database, the musician table contains the following data:
>
> ```
> musician_name     phone         musician_type
> ----------------  ------------  -------------
> Diva DeLuca       615-758-7836  Opera Singer
> Skeeter Sullivan  629-209-2332  Bluegrass Singer
> Tex Macaroni      915-789-1721  Country Singer
> Bronzy Bohannon   212-211-1216  Sax Player
> ```
>
> Write a query that finds all the singers from Nashville, Tennessee. Nashville has two area codes, 615 and 629, so you'll want to find phone numbers that start with those numbers. You can find singers by finding the text `Singer` somewhere in the `musician_type` column.

not like

The `not like` operator can be used to find strings that do *not* match some pattern. It also uses the `%` and `_` wildcard characters. For example, to reverse your logic for the `like` example, enter the following:

```
select  *
from    three_letter_term
where   term not like '_at';
```

The results are words in the `three_letter_term` table that do not end in *at*:

```
term
----
dog
egg
ape
```

Similarly, you can find billionaires whose last names do not start with the letter *M* using this query:

```
select  *
from    billionaire
where   last_name not like 'M%';
```

The results might look like this:

```
first_name   last_name
----------   ---------
Jeff         Bezos
Bill         Gates
Mark         Zuckerberg
```

exists

The exists operator checks to see if a subquery returns at least one row. Here you go back to the customer table in the not between example and use exists to see whether the table has at least one millennial:

```
select 'There is at least one millennial in this table'
where exists
(
    select  *
    from    customer
    where   birthyear between 1981 and 1996
);
```

There are millennials in the customer table, so your result is:

```
There is at least one millennial in this table
```

If there had been no customers born between 1981 and 1996, your query wouldn't have returned any rows, and the text There is at least one millennial in this table would not have been shown.

You might see the same query written using select 1 instead of select * in the subquery:

```
select 'There is at least one millennial in this table'
where exists
(
    select  1
    from    customer
    where   birthyear between 1981 and 1996
);
```

In this query, it doesn't matter if you select * or 1 because you're looking for at least one customer that matches your description. All you really care about is that the inner query returned *something*.

Checking Booleans

In Chapter 4, you learned that booleans can have one of two values: `true` or `false`. You can use special syntax, `is true` or `is false`, to return only results with one value or the other. In this example, you return a list of employed bachelors in the `bachelor` table by using the `is true` syntax in the `employed_flag` column:

```
select  *
from    bachelor
where   employed_flag is true;
```

This query causes MySQL to return only rows for bachelors who are employed.

To check bachelors whose `employed_flag` value is set to `false`, use `is false`:

```
select  *
from    bachelor
where   employed_flag is false;
```

Now MySQL returns only rows for bachelors who are unemployed.

You can check the value of boolean columns in other ways as well. These lines are all equivalent ways of checking for `true` values:

```
employed_flag is true
employed_flag
employed_flag = true
employed_flag != false
employed_flag = 1
employed_flag != 0
```

The following lines are all equivalent ways to check for `false` values:

```
employed_flag is false
not employed_flag
employed_flag = false
employed_flag != true
employed_flag = 0
employed_flag != 1
```

As you can see here, a value of `1` is equivalent to `true` and a value of `0` is equivalent to `false`.

or Conditions

You can use MySQL's `or` keyword to check that at least one of two conditions has been met.

Consider this table called applicant, which contains information about job applicants.

```
name          associates_degree_flag  bachelors_degree_flag  years_experience
------------  ----------------------  ---------------------  ----------------
Joe Smith                0                      1                    7
Linda Jones              1                      0                    2
Bill Wang                0                      1                    1
Sally Gooden             1                      0                    0
Katy Daly                0                      0                    0
```

The associates_degree_flag and bachelors_degree_flag columns are booleans, where 0 represents false and 1 represents true.

In the following query, you select from the applicant table to get a list of qualified applicants for a job that requires a bachelor's degree *or* two or more years of experience:

```
select  *
from    applicant
where   bachelors_degree_flag is true
or      years_experience >= 2;
```

The results are:

```
name          associates_degree_flag  bachelors_degree_flag  years_experience
------------  ----------------------  ---------------------  ----------------
Joe Smith                0                      1                    7
Linda Jones              1                      0                    2
Bill Wang                0                      1                    1
```

Say you need to write a query with both the and (both conditions must be met) and or (either condition must be met) keywords. In this case, you can use parentheses to group your conditions so that MySQL will return the correct results.

Let's see how using parentheses can be beneficial. Here you create another query with the applicant table for a new job that requires applicants to have two or more years' experience *and* either an associate's degree *or* a bachelor's degree:

```
select  *
from    applicant
where   years_experience >= 2
and     associates_degree_flag is true
or      bachelors_degree_flag is true;
```

The results of this query are not what you expected:

```
name          associates_degree_flag  bachelors_degree_flag  years_experience
------------  ----------------------  ---------------------  ----------------
Joe Smith                0                      1                    7
Linda Jones              1                      0                    2
Bill Wang                0                      1                    1
```

Bill doesn't have two or more years' experience, so why did he appear in your result set?

The query uses both an `and` and an `or`. The `and` has a higher *operator precedence* than the `or`, which means `and` gets evaluated before `or`. This caused your query to find applicants that met at least one of the following two conditions:

- Two or more years' experience *and* an associate's degree

or

- A bachelor's degree

That's not what you intended when you wrote the query. You can correct the problem by using parentheses to group your conditions:

```
select  *
from    applicant
where   years_experience >= 2
and     (
        associates_degree_flag is true
or      bachelors_degree_flag is true
        );
```

Now the query finds applicants that meet these conditions:

- Two or more years' experience

and

- An associate's degree *or* a bachelor's degree

Your results should now be in line with your expectations:

```
name          associates_degree_flag  bachelors_degree_flag  years_experience
------------  ----------------------  ---------------------  ----------------
Joe Smith                0                      1                       7
Linda Jones              1                      0                       2
```

NOTE *Even in cases where parentheses won't change the results returned by your query, using them is a best practice because it makes your code more readable.*

TRY IT YOURSELF

7-2. In the `airport` database, the `boarding` table contains the following data:

```
passenger_name  license_flag  student_id_flag  soc_sec_card_flag
--------------  ------------  ---------------  -----------------
Frank Flyer          1              0                 0
Rhonda Runway        0              0                 1
Sam Suitcase         0              1                 1
Pam Prepared         1              1                 1
```

(continued)

In order to board a flight, a passenger must have a license, along with either a student ID or a Social Security card. The following query was written to identify passengers who are allowed to board:

```
select  *
from    boarding
where   license_flag is true
and     student_id_flag is true
or      soc_sec_card_flag is true;
```

But it isn't returning the correct results:

```
passenger_name  license_flag  student_id_flag  soc_sec_card_flag
--------------  ------------  ---------------  -----------------
Rhonda Runway        0               0                 1
Sam Suitcase         0               1                 1
Pam Prepared         1               1                 1
```

Only `Pam Prepared` should be appearing in this list. How would you change the query to get the correct results?

Summary

In this chapter, you learned various ways to compare values in MySQL through comparison operators, such as checking whether values are equal, null, or within a range, or if they match a pattern. You also learned how to check that at least one condition is met in your queries.

In the next chapter, you'll take a look at using MySQL's built-in functions, including those that deal with mathematics, dates, and strings. You'll also learn about aggregate functions and how to use them for groups of values.

8

CALLING BUILT-IN MYSQL FUNCTIONS

MySQL has hundreds of prewritten functions that perform a variety of tasks. In this chapter, you'll review some common functions and learn how to call them from your queries. You'll work with aggregate functions, which return a single value summary based on many rows of data in the database, and functions that help perform mathematical calculations, process strings, deal with dates, and much more.

In Chapter 11, you'll learn to create your own functions, but for now you'll focus on calling MySQL's most useful built-in functions. For an up-to-date list of all the built-in functions, the best source is the MySQL reference manual. Search online for "MySQL built-in function and operator reference," and bookmark the web page in your browser.

What Is a Function?

A *function* is a set of saved SQL statements that performs some task and returns a value. For example, the pi() function determines the value of pi and returns it. Here's a simple query that calls the pi() function:

```
select pi();
```

Most of the queries you've seen thus far include a from clause that specifies which table to use. In this query, you aren't selecting from any table, so you can call the function without from. It returns the following result:

```
pi()
----------
3.141593
```

For common tasks such as this, it makes more sense to use MySQL's built-in function rather than having to remember the value every time you need it.

Passing Arguments to a Function

As you just saw, functions return a value. Some functions also let you pass values to them. When you call the function, you can specify a value that it should use. The values you pass to a function are called *arguments*.

To see how arguments work, you'll call the upper() function, which allows you to accept one argument: a string value. The function determines what the uppercase equivalent of that string is and returns it. The following query calls upper() and specifies an argument of the text rofl:

```
select upper('rofl');
```

The result is as follows:

```
upper('rofl')
--------------
ROFL
```

The function translated each letter to uppercase and returned ROFL.

TRY IT YOURSELF

8-1. You can use the lower() function, which takes one argument, to return the lowercase version of a string. Call the lower() function with the argument E.E. Cummings and see what it returns.

8-2. The now() function returns the current date and time. Call now() with no arguments.

In some functions, you can specify more than one argument. For example, `datediff()` allows you to specify two dates as arguments and then returns the difference in days between them. Here you call `datediff()` to find out how many days there are between Christmas and Thanksgiving in 2024:

```
select datediff('2024-12-25', '2024-11-28');
```

The result is:

```
datediff('2024-12-25', '2024-11-28')
27
```

When you called the `datediff()` function, you specified two arguments, the date of Christmas and the date of Thanksgiving, and separated them by commas. The function calculated the difference in days and returned that value (`27`).

Functions accept different numbers and types of values. For example, `upper()` accepts one string value, while `datediff()` accepts two `date` values. As you'll see in this chapter, other functions accept values that are an integer, a boolean, or another data type.

Optional Arguments

Some functions accept an optional argument, in which you can supply another value for a more specific result when you call the function. The `round()` function, for example, which rounds decimal numbers, accepts one argument that must be provided and a second argument that is optional. If you call `round()` with the number you want rounded as the only argument, it will round the number to zero places. Try calling the `round()` function with one argument of `2.71828`:

```
select round(2.71828);
```

The `round()` function returns your rounded number with zero digits after the decimal point, which also removes the decimal point itself:

```
round(2.71828)
--------------
      3
```

If you supply `round()` with its optional argument, you can specify how many places after the decimal point you want it to round. Try calling `round()` with a first argument of `2.71828` and a second argument of `2`, separating the arguments with a comma:

```
select round(2.71828, 2);
```

Now the result is:

```
round(2.71828)
--------------
     2.72
```

This time, round() returns a rounded number with two digits after the decimal point.

GETTING HELP

MySQL provides the help statement so you can get help from the MySQL reference manual. If you type help round, for example, MySQL provides a wealth of information about the round() function, including the URL of the manual page for it and examples:

```
> help round
Name: 'ROUND'
Description:
Syntax:
ROUND(X), ROUND(X,D)

Rounds the argument X to D decimal places. The rounding algorithm
depends on the data type of X. D defaults to 0 if not specified. D can
be negative to cause D digits left of the decimal point of the value X
to become zero. The maximum absolute value for D is 30; any digits in
excess of 30 (or -30) are truncated.

URL: https://dev.mysql.com/doc/refman/8.0/en/mathematical-functions.html

Examples:
mysql> SELECT ROUND(-1.23);
        -> -1
mysql> SELECT ROUND(-1.58);
        -> -2
mysql> SELECT ROUND(1.58);
        -> 2
mysql> SELECT ROUND(1.298, 1);
        -> 1.3
mysql> SELECT ROUND(1.298, 0);
        -> 1
mysql> SELECT ROUND(23.298, -1);
        -> 20
mysql> SELECT ROUND(.12345678901234567890123456789012345, 35);
        -> 0.123456789012345678901234567890
```

The help statement can also assist with topics other than functions. For example, you could type help 'data types' to learn about MySQL data types:

```
> help 'data types'
You asked for help about help category: "Data Types"
For more information, type 'help <item>', where <item> is one of the
following
topics:
   AUTO_INCREMENT
   BIGINT
   BINARY
   BIT
   BLOB
   BLOB DATA TYPE
   BOOLEAN
   CHAR
   CHAR BYTE
   DATE
   DATETIME
   --snip--
```

The help statement is case-insensitive, so getting help on round and ROUND will return the same information.

Calling Functions Within Functions

You can use the results of one function in a call to another function by wrapping, or nesting, functions.

Say you want to get the rounded value of pi. You can wrap your call to the pi() function within a call to the round() function:

```
select round(pi());
```

The result is:

```
round(pi())
-----------
     3
```

The innermost function gets executed first and the results are passed to the outer function. The call to the pi() function returns 3.141593, and that value is passed as an argument to the round() function, which returns 3.

NOTE *If you find queries with nested functions hard to read, you can format your SQL to put the inner functions on their own line and indent them as follows:*

```
select round(
         pi()
       );
```

You can modify your query and round pi to two digits by specifying a value in the round() function's optional second argument, like so:

```
select round(pi(), 2);
```

The result is:

```
round(pi(), 2)
-------------
     3.14
```

This call to the pi() function returns 3.141593, which is passed to round() as the function's first argument. The statement evaluates to round(3.141593,2), which returns 3.14.

Calling Functions from Different Parts of Your Query

You can call functions in the select list of your query and also in the where clause. For example, take a look at the movie table, which contains the following data about movies:

```
movie_name        star_rating  release_date
----------------- -----------  ------------
Exciting Thriller 4.72         2024-09-27
Bad Comedy        1.2          2025-01-02
OK Horror         3.1789       2024-10-01
```

The star_rating column holds the average number of stars that viewers rated the movie on a scale of 1 to 5. You've been asked to write a query to display movies that have more than 3 stars and a release date in 2024. You also need to display the movie name in uppercase and round the star rating:

```
select upper(movie_name),
       round(star_rating)
from   movie
where  star_rating > 3
and    year(release_date) = 2024;
```

First, you use the upper() and round() functions in the select list of the query. You wrap the movie name values in the upper() function and wrap the star rating value in the round() function. You then specify that you're pulling data from the movie table.

In the where clause, you call the year() function and specify one argument: the release_date from the movie table. The year() function returns the year of the movie's release, which you compare (=) to 2024 to display only movies with a release date in 2024.

The results are:

```
upper(movie_name) round(star_rating)
----------------- ------------------
EXCITING THRILLER         5
OK HORROR                 3
```

Aggregate Functions

An *aggregate* function is a type of function that returns a single value based on multiple values in the database. Common aggregate functions include `count()`, `max()`, `min()`, `sum()`, and `avg()`. In this section, you'll see how to call these functions with the following `continent` table:

```
continent_id  continent_name  population
------------  --------------  ----------
1             Asia            4641054775
2             Africa          1340598147
3             Europe          747636026
4             North America   592072212
5             South America   430759766
6             Australia       43111704
7             Antarctica      0
```

count()

The `count()` function returns the number of rows returned from a query, and can help answer questions about your data like "How many customers do you have?" or "How many complaints did you get this year?"

You can use the `count()` function to determine how many rows are in the `continent` table, like so:

```
select  count(*)
from    continent;
```

When you call the `count()` function, you use an asterisk (or a wildcard) between the parentheses to count all rows. The asterisk selects all rows from a table, including all of each row's column values.

The result is:

```
count(*)
--------
   7
```

Use a `where` clause to select all continents with a population of more than 1 billion:

```
select  count(*)
from    continent
where   population > 1000000000;
```

The result is:

```
count(*)
--------
   2
```

The query returns `2` because only two continents, Asia and Africa, have more than 1 billion people.

max()

The max() function returns the maximum value in a set of values, and can help answer questions like "What was the highest yearly inflation rate?" or "Which salesperson sold the most cars this month?"

Here you use the max() function to find the maximum population for any continent in the table:

```
select max(population)
from   continent;
```

The result is:

```
max(population)
---------------
  4641054775
```

When you call the max() function, it returns the number of people who live in the most populated continent. The row in the table with the highest population for any continent is Asia, with a population of 4,641,054,775.

Aggregate functions like max() can be particularly useful in subqueries. Step away from the continent table for a moment, and turn your attention to the train table:

```
train            mile
---------------  ----
The Chief        8000
Flying Scotsman  6500
Golden Arrow     2133
```

Here you'll use max() to help determine which train in the train table has traveled the most miles:

```
select   *
from     train
where    mile =
(
  select max(mile)
  from   train
);
```

In the inner query, you select the maximum number of miles that any train in your table has traveled. In the outer query, you display all the columns for trains that have traveled that number of miles.

The result is:

```
train_name  mile
----------  ----
The Chief   8000
```

min()

The min() function returns the minimum value in a set of values, and can help answer questions such as "What is the cheapest price for gas in town?" or "Which metal has the lowest melting point?"

Let's return to the continent table. Use the min() function to find the population of the least populated continent:

```
select min(population)
from   continent;
```

When you call the min() function, it returns the minimum population value in the table:

```
min(population)
---------------
       0
```

The row in the table with the lowest population is Antarctica, with 0.

sum()

The sum() function calculates the sum of a set of numbers, and helps answer questions like "How many bikes are there in China?" or "What were your total sales this year?"

Use the sum() function to get the total population of all the continents, like so:

```
select sum(population)
from   continent;
```

When you call the sum() function, it returns the sum total of the population for every continent.

The result is:

```
max(population)
---------------
   7795232630
```

avg()

The avg() function returns the average value based on a set of numbers, and can help answer questions including "What is the average amount of snow in Wisconsin?" or "What is the average salary for a doctor?"

Use the avg() function to find the average population of the continents:

```
select avg(population)
from   continent;
```

When you call the avg() function, it returns the average population value of the continents in the table:

```
avg(population)
---------------
1113604661.4286
```

MySQL arrives at 1,113,604,661.4286 by totaling the population of every continent (7,795,232,630) and dividing that result by the number of continents (7).

Now, use the avg() function in a subquery to display all continents that are less populated than the average continent:

```
select   *
from     continent
where    population <
(
  select avg(population)
  from   continent
);
```

The inner query selects the average population size for all of continents: 1,113,604,661.4286 people. The outer query selects all columns from the continent table for continents with populations less than that value.

The result is:

```
continent_id  continent_name  population
------------  --------------  ----------
     3        Europe           747636026
     4        North America    592072212
     5        South America    430759766
     6        Australia         43111704
     7        Antarctica               0
```

TRY IT YOURSELF

In the music database, the genre_stream table contains the following data:

```
Genre              Stream
-----------------  -------
R&B, Hip Hop       3102456
Rock               1577569
Pop                1298756
Country             764789
Latin               601758
Dance, Electronic   308745
```

8-3. Write a query to determine how many rows are in the table.

8-4. Write a query to find the average number of streams for all genres.

group by

A `group by` clause tells MySQL how you want your results grouped, and can be used only in queries with aggregate functions. To see how `group by` works, take a look at the `sale` table, which stores a company's sales:

```
sale_id  customer_name  salesperson  amount
-------  -------------  -----------  ------
1        Bill McKenna   Sally         12.34
2        Carlos Souza   Sally         28.28
3        Bill McKenna   Tom            9.72
4        Bill McKenna   Sally         17.54
5        Jane Bird      Tom           34.44
```

You can use the `sum()` aggregate function to add the sales amounts, but do you want to calculate one grand total for all sales, sum the amounts by customer, sum the amounts by salesperson, or calculate the totals that each salesperson sold to each customer?

To display amounts summed by customer, you `group by` the `customer_name` column, as in Listing 8-1.

```
select sum(amount)
from   sale
group by customer_name;
```

Listing 8-1: A query to sum amounts by customer

The results are as follows:

```
sum(amount)
-----------
      39.60
      28.28
      34.44
```

The sum total of the amount spent by customer Bill McKenna is $39.60; for Carlos Souza, it's $28.28; and for Jane Bird, it's $34.44. The results are ordered alphabetically by the customer's first name.

Alternatively, you may want to see sum totals of the amounts by salesperson. Listing 8-2 shows you how to use `group by` on the `salesperson_name` column.

```
select sum(amount)
from   sale
group by salesperson_name;
```

Listing 8-2: A query to sum amounts by salesperson

Your results are:

```
sum(amount)
-----------
      58.16
      44.16
```

The total amount sold by Sally is $58.16, and for Tom it's $44.16.

Because `sum()` is an aggregate function, it can operate on any number of rows and will return one value. The `group by` statement tells MySQL which rows you want `sum()` to operate on, so the syntax `group by salesperson_name` sums up the amounts for each salesperson.

Now say that you want to see just one row with a sum of every `amount` in the table. In this case, you don't need to use `group by`, since you aren't summing up by any group. Your query should look like the following:

```
select  sum(amount)
from    sale;
```

The result should be:

```
sum(amount)
-----------
     102.32
```

The `group by` clause works with all aggregate functions. For example, you could use `group by` with `count()` to return the count of sales for each salesperson, as in Listing 8-3.

```
select count(*)
from   sale
group by salesperson_name;
```

Listing 8-3: A query to count rows for each salesperson

The result is:

```
count(*)
--------
   3
   2
```

The query counted three rows in the `sales` table for Sally and two rows for Tom.

Or you can use `avg()` to get the average sale amount and group by `salesperson_name` to return the average sale amount per salesperson, as shown in Listing 8-4.

```
select   avg(amount)
from     sale
group by salesperson_name;
```

Listing 8-4: A query to get the average amount sold by each salesperson

The result is:

```
avg(amount)
-----------
  19.386667
  22.080000
```

The results show that the average amount of each sale for Sally was $19.386667, and the average amount of each sale for Tom was $22.08.

When looking at these results, however, it's not immediately clear which salesperson's average was $19.386667 and which salesperson's was $22.08. To clarify that, let's modify the query to display more information in the result set. In Listing 8-5, you select the salesperson's name as well.

```
select   salesperson_name,
         avg(amount)
from     sale
group by salesperson_name;
```

Listing 8-5: A query to display the salesperson's name and their average amount sold

The results of your modified query are:

```
salesperson_name  avg(amount)
----------------  -----------
Sally               19.386667
Tom                 22.080000
```

Your averages appear with the same values, but now the salesperson's name appears next to them. Adding this extra information makes your results much easier to understand.

After you've written several queries that use aggregate functions and `group by`, you might notice that you usually group by the same columns that you selected in the query. For example, in Listing 8-5, you selected the `salesperson_name` column and also grouped by the `salesperson_name` column.

To help you determine which column(s) to group by, look at the *select list*, or the part of the query between the words `select` and `from`. The select list contains the items you want to select from the database table; you almost always want to group by this same list. The only part of the select list that shouldn't be part of the `group by` statement are the aggregate functions called.

For example, take a look at this `theme_park` table, which contains data from six different theme parks, including their country, state, and the city where they are located:

```
country  state           city                park
-------  -----------     ------------------  -----------------
USA      Florida         Orlando             Disney World
USA      Florida         Orlando             Universal Studios
USA      Florida         Orlando             SeaWorld
USA      Florida         Tampa               Busch Gardens
Brazil   Santa Catarina  Balneario Camboriu  Unipraias Park
Brazil   Santa Catarina  Florianopolis       Show Water Park
```

Say you want to select the country, state, and the number of parks for those countries and states. You might start to write your SQL statement like this:

```
select country,
       state,
       count(*)
from   theme_park;
```

This query is incomplete, however, and running it will return an error message or incorrect results, depending on your configuration settings.

You should group by everything you've selected *that is not an aggregate function*. In this query, the columns you've selected, country and state, are not aggregate functions, so you will use group by with them:

```
select   country,
         state,
         count(*)
from     theme_park
group by country,
         state;
```

The results are as follows:

```
country state         count(*)
------  ------------- --------
USA     Florida          4
Brazil  Santa Catarina   2
```

As you can see, the query now returns the correct results.

TRY IT YOURSELF

8-5. You can find the theme_park table in the vacation database. Write a query to select the country and the count of parks in each country. Do not display the state or the city. Which column should you group by in your query?

String Functions

MySQL provides several functions to help you work with character strings and perform tasks such as comparing, formatting, and combining strings. Let's take a look at the most useful string functions.

concat()

The concat() function *concatenates*, or joins, two or more strings together. For example, say you have the following phone_book table:

```
first_name  last_name
----------  ----------
Jennifer    Perez
Richard     Johnson
John        Moore
```

You can write a query to display first and last names together, separated by a space character:

```
select  concat(first_name, ' ', last_name)
from    phone_book;
```

The results should be as follows:

```
concat(first_name, ' ', last_name)
----------------------------------
Jennifer Perez
Richard Johnson
John Moore
```

The names appear as one string, separated by a space.

format()

The `format()` function formats a number by adding commas and showing the requested number of decimal points. For example, let's revisit the `continent` table and select the population of Asia as follows:

```
select  population
from    continent
where   continent_name = 'Asia';
```

The result is:

```
population
----------
4641054775
```

It's difficult to tell whether the population of Asia is about 4.6 billion or 464,000,000. To make the results more readable, you can format the `population` column with commas using the `format()` function like so:

```
select format(population, 0)
from   continent;
```

The `format()` function takes two arguments: a number to format and the number of positions to show after the decimal point. You called `format()` with two arguments: the `population` column and the number `0`.

NOTE *The `format()` function requires two arguments, meaning that you need to specify `0` as the second argument if you don't want your result to show any numbers after the decimal point. This differs from the `round()` function, which allows you to leave the second argument blank. If you were to leave the second argument in `format()` blank, you'd get an error.*

Now that the population column has been formatted with commas, it's clear in the result that Asia has around 4.6 billion people:

```
population
-------------
4,641,054,775
```

Now call the format() function to format the number 1234567.89 with five digits after the decimal point:

```
select format(1234567.89, 5);
```

The result is:

```
format(1234567.89, 5)
---------------------
    1,234,567.89000
```

The format() function accepts 1234567.89 as the number to be formatted in the first argument, adds commas, and add trailing zeros so that the result is displayed with five decimal positions.

left()

The left() function returns some number of characters from the left side of a value. Consider the following taxpayer table:

```
last_name  soc_sec_no
---------  ------------
Jagger     478-555-7598
McCartney  478-555-1974
Hendrix    478-555-3555
```

To select last names from the taxpayer table, and also select the first three characters of the last_name column, you can write the following:

```
select  last_name,
        left(last_name, 3)
from    taxpayer;
```

The result is:

```
last_name   left(last_name, 3)
----------  -----------------
Jagger      Jag
McCartney   McC
Hendrix     Hen
```

The left() function is helpful in cases when you want to disregard the characters on the right.

right()

The `right()` function returns some number of characters from the right side of a value. Continue using the `taxpayer` table to select the last four digits of the taxpayers' Social Security numbers:

```
select  right(soc_sec_no, 4)
from    taxpayer;
```

The result is:

```
right(soc_sec_no, 4)
--------------------
        7598
        1974
        3555
```

The `right()` function selects the rightmost characters without the characters on the left.

lower()

The `lower()` function returns the lowercase version of a string. Select the taxpayers' last names in lowercase:

```
select  lower(last_name)
from    taxpayer;
```

The result is:

```
lower(last_name)
----------------
jagger
mccartney
hendrix
```

upper()

The `upper()` function returns the uppercase version of a string. Select the taxpayers' last names in uppercase:

```
select  upper(last_name)
from    taxpayer;
```

The result is:

```
upper(last_name)
----------------
JAGGER
MCCARTNEY
HENDRIX
```

> **THE POWER OF COMBINING FUNCTIONS**
>
> MySQL functions can be most useful when you combine them. Say you work for a tax agency, and you're tasked with creating a new taxpayer identifier that will comprise the first three characters of the taxpayer's last name in uppercase letters concatenated with the last four digits of their Social Security number. You can use the `concat()`, `upper()`, `left()`, and `right()` functions together to build the new taxpayer ID, like so:
>
> ```
> select  last_name,
>         concat(
>           upper(
>             left(last_name, 3)
>           ),
>           right(soc_sec_no, 4)
>         ) as new_taxpayer_id
> from    taxpayer;
> ```
>
> The result is:
>
> ```
> last_name  new_taxpayer_id
> ---------  ---------------
> Jagger        JAG7598
> McCartney     MCC1974
> Hendrix       HEN3555
> ```
>
> The result set shows two columns: the `last_name` column that you selected directly from the `taxpayer` table, and the `new_taxpayer_id` column that you built using parts of the `last_name` and `soc_sec_no` columns and some built-in functions.
>
> To get the `new_taxpayer_id` values, you used the `left()` function to get the first three characters from the `last_name` column; called the `upper()` function to convert those three characters to uppercase; and used the `right()` function to get the last four digits from the `soc_sec_no` column. Then you passed those values into the `concat()` function to combine them into one string. Finally, you used the `as new_taxpayer_id` syntax to create a column alias of `new_taxpayer_id` so that the second column appears with that heading.

substring()

The `substring()` function returns part of a string and takes three arguments: a string, the starting character position of the substring you want, and the ending character position of the substring you want.

You can extract the substring `gum` from the string `gumbo` by using this query:

```
select substring('gumbo', 1, 3);
```

The result is:

```
substring('gumbo', 1, 3)
------------------------
          gum
```

In gumbo, g is the first character, u is the second character, and m is the third. Selecting a substring starting at character 1 and going to character 3 returns those first three characters.

The second argument to the substring() function can accept a negative number. If you pass a negative number to it, the beginning position of your substring will be calculated by counting backward from the end of the string. For example:

```
select substring('gumbo', -3, 2);
```

The result is:

```
substring('gumbo', -3, 2)
------------------------
          mb
```

The string gumbo has five characters. You asked substring() to start your substring at the end of the string minus three character positions, which is position 3. Your third argument was 2, so your substring will start at the third character 3 and go for two characters, yielding the mb substring.

The third argument to the substring() function is optional. You can provide just the first two arguments—a string and the starting character position—to return the set of characters between the starting position until the end of the string:

```
select substring('MySQL', 3);
```

The result is:

```
substring('MySQL', 3)
------------------------
          SQL
```

The substring() function returned all the characters starting at the third character of the string MySQL, going all the way to the end of the string, resulting in SQL.

MySQL provides an alternate syntax for substring() that uses the from and for keywords. For example, to select the first three characters of the word gumbo, use the following syntax:

```
select substring('gumbo' from 1 for 3);
```

This substring starts at the first character and continues for three characters. The result is as follows:

```
substring('gumbo' from 1 for 3)
------------------------------
            gum
```

This result is the same as the first substring example you saw, but you might find this syntax easier to read.

NOTE *The* `substring()` *function has a synonym called* `substr()`*. Calling* `substring('MySQL', 3)` *and* `substr('MySQL', 3)` *yields identical results.*

trim()

The `trim()` function strips any number of leading or trailing characters from a string. You can specify the characters you want removed, as well as whether you want the leading characters removed, the trailing characters removed, or both.

For example, if you have the string `**instructions**`, you could use `trim()` to return the string with the asterisks removed like so:

```
select trim(leading  '*' from '**instructions**') as column1,
       trim(trailing '*' from '**instructions**') as column2,
       trim(both     '*' from '**instructions**') as column3,
       trim(         '*' from '**instructions**') as column4;
```

In `column1`, you trim the `leading` asterisks. In `column2`, you trim the `trailing` asterisks. In `column3`, you trim `both` the leading and trailing asterisks. When you don't specify `leading`, `trailing`, or `both`, as in `column4`, MySQL defaults to trimming both.

The results are as follows:

```
column1          column2          column3        column4
--------------   --------------   ------------   ------------
instructions**   **instructions   instructions   instructions
```

By default, `trim()` removes space characters. This means that if you have space characters around a string, you can use `trim()` without having to specify the character you want to strip:

```
select trim('   asteroid   ');
```

The result is the string `asteroid` with no spaces on either side:

```
trim('   asteroid   ')
----------------------
asteroid
```

The `trim()` function removes spaces from both sides of a string by default.

ltrim()

The `ltrim()` function removes leading spaces from the left side of a string:

```
select ltrim('   asteroid   ');
```

The result is the string `asteroid` with no spaces on the left side of it:

```
ltrim('   asteroid   ')
----------------------
asteroid
```

The spaces to the right are unaffected.

rtrim()

The `rtrim()` function removes trailing spaces from the right side of a string:

```
select rtrim('   asteroid   ');
```

The result is the string `asteroid` with no spaces on the right side of it:

```
rtrim('   asteroid   ')
----------------------
   asteroid
```

The spaces to the left are unaffected.

TRY IT YOURSELF

8-6. A zip code (such as 24701 or 79936) is used for US postal delivery. Each character in a zip code has meaning. The first character is the national area. The second and third characters represent the sectional center. The fourth and fifth characters are for the associate post office.

The `mail` database contains an `address` table that has a `zip_code` column. The table contains character strings for the following zip codes: `94103`, `37188`, and `96718`.

Write a query that selects the zip code and uses the `substring()` function to select the other parts of the zip code from the `zip_code` column. The query should produce the following output:

```
zip_code  national_area  sectional_center  associate_post_office
--------  -------------  ----------------  ---------------------
94103                 9                41                     03
37188                 3                71                     88
96718                 9                67                     18
```

Date and Time Functions

MySQL provides date-related functions that help you perform tasks like getting the current date and time, selecting a part of the date, and calculating how many days there are between two dates.

As you saw in Chapter 4, MySQL provides the `date`, `time`, and `datetime` data types, where `date` contains a month, day, and year; `time` contains hours, minutes, and seconds; and `datetime` has all of those parts because it comprises both a date and a time. These are the formats MySQL uses to return many of the results of the functions you'll see here.

curdate()

The `curdate()` function returns the current date in the `date` format:

```
select curdate();
```

Your result should look similar to the following:

```
curdate()
----------
2024-12-14
```

Both `current_date()` and `current_date` are synonyms for `curdate()` and will produce identical results.

curtime()

The `curtime()` function returns the current time in the `time` format:

```
select curtime();
```

Your result should look similar to the following:

```
curtime()
---------
09:02:41
```

For me, the current time is 9:02 AM and 41 seconds. Both `current_time()` and `current_time` are synonyms for `curtime()` and will produce identical results.

now()

The `now()` function returns the current date and time in a `datetime` format:

```
select now();
```

Your results should look similar to the following:

```
now()
-------------------
2024-12-14 09:02:18
```

Both current_timestamp() and current_timestamp are synonyms for now() and will produce identical results.

date_add()

The date_add() function adds some amount of time to a date value. To add (or subtract) from date values, you use an *interval*, a value that you can use to perform calculations on dates and times. With an interval, you can supply a number and a unit of time, like 5 day, 4 hour, or 2 week. Consider the following table called event:

```
event_id  eclipse_datetime
--------  -------------------
   374    2024-10-25 11:01:20
```

To select the eclipse_datetime date from the event table and add 5 days, 4 hours, and 2 weeks to the date, you use date_add() with interval as follows:

```
select  eclipse_datetime,
        date_add(eclipse_datetime, interval 5 day)  as add_5_days,
        date_add(eclipse_datetime, interval 4 hour) as add_4_hours,
        date_add(eclipse_datetime, interval 2 week) as add_2_weeks
from    event
where   event_id = 374;
```

NOTE *The units of time are singular. For example, you use* minute*, not* minutes*. Other units of time commonly used with intervals include second, month, and year.*

Your results should look similar to this:

```
eclipse_datetime     add_5_days           add_4_hours          add_2_weeks
-------------------  -------------------  -------------------  -------------------
2024-10-25 11:01:20  2024-10-30 11:01:20  2024-10-25 15:01:20  2024-11-08 11:01:20
```

The results show that the intervals of 5 days, 4 hours, and 2 weeks were added to the eclipse date and time and have been listed in the columns you specified.

date_sub()

The date_sub() function subtracts a time interval from a date value. For example, here you subtract the same time intervals in the previous example from the eclipse_datetime column of the event table:

```
select  eclipse_datetime,
        date_sub(eclipse_datetime, interval 5 day)  as sub_5_days,
        date_sub(eclipse_datetime, interval 4 hour) as sub_4_hours,
        date_sub(eclipse_datetime, interval 2 week) as sub_2_weeks
from    event
where   event_id = 374;
```

The results are:

```
eclipse_datetime     sub_5_days           sub_4_hours          sub_2_weeks
-------------------  -------------------  -------------------  -------------------
2024-10-25 11:01:20  2024-10-20 11:01:20  2024-10-25 07:01:20  2024-10-11 11:01:20
```

The results show that the intervals of 5 days, 4 hours, and 2 weeks were subtracted from the eclipse date and time and have been listed in the columns you specified.

extract()

The `extract()` function pulls out specified parts of a `date` or a `datetime` value. It uses the same units of time as `date_add()` and `date_sub()`, like `day`, `hour`, and `week`.

In this example, you select some parts of your `eclipse_datetime` column:

```
select  eclipse_datetime,
        extract(year from eclipse_datetime)   as year,
        extract(month from eclipse_datetime)  as month,
        extract(day from eclipse_datetime)    as day,
        extract(week from eclipse_datetime)   as week,
        extract(second from eclipse_datetime) as second
from    event
where   event_id = 374;
```

The `extract()` function takes the `eclipse_datetime` value from the `event` table and displays the individual parts requested by the column names you specify. The results are as follows:

```
eclipse_datetime     year  month  day  week  second
-------------------  ----  -----  ---  ----  ------
2024-10-25 11:01:20  2024     10   25    43      20
```

MySQL provides other functions you can use for the same purpose as `extract()`, including `year()`, `month()`, `day()`, `week()`, `hour()`, `minute()`, and `second()`. This query achieves the same result as the preceding one:

```
select  eclipse_datetime,
        year(eclipse_datetime)   as year,
        month(eclipse_datetime)  as month,
        day(eclipse_datetime)    as day,
        week(eclipse_datetime)   as week,
        second(eclipse_datetime) as second
from    event
where   event_id = 374;
```

You can also use the `date()` and `time()` functions to select just the `date` or `time` portion of a `datetime` value:

```
select  eclipse_datetime,
        date(eclipse_datetime)   as date,
        time(eclipse_datetime)   as time
from    event
where   event_id = 374;
```

The results are:

```
eclipse_datetime     date        time
-------------------  ----------  --------
2024-10-25 11:01:20  2024-10-25  11:01:20
```

As you can see, the `date()` and `time()` functions provide a quick way to extract just the date or the time from a `datetime` value.

datediff()

The `datediff()` function returns the number of days between two dates. Say you want to check how many days there are between New Year's Day and Cinco de Mayo in 2024:

```
select datediff('2024-05-05', '2024-01-01');
```

The result is 125 days:

```
datediff('2024-05-05', '2024-01-01')
------------------------------------
                  125
```

If the date argument on the left is more recent than the date argument on the right, `datediff()` will return a positive number. If the date on the right is more recent, `datediff()` will return a negative number. If the two dates are the same, `0` will be returned.

date_format()

The `date_format()` function formats a date according to a format string that you specify. The format string is made up of characters that you add and *specifiers* that start with a percent sign. The most common specifiers are listed in Table 8-1.

Table 8-1: Common Specifiers

Specifier	Description
%a	Abbreviated weekday name (Sun–Sat)
%b	Abbreviated month name (Jan–Dec)
%c	Month, numeric (1–12)
%D	Day of the month with suffix (1st, 2nd, 3rd, . . .)
%d	Day of the month, two digits with a leading zero where applicable (01–31)
%e	Day of the month (1–31)
%H	Hour with leading zero where applicable (00–23)
%h	Hour (01–12)
%i	Minutes (00–59)
%k	Hour (0–23)

(continued)

Table 8-1: Common Specifiers *(continued)*

Specifier	Description
%l	Hour (1–12)
%M	Month name (January–December)
%m	Month (00–12)
%p	AM or PM
%r	Time, 12-hour (hh:mm:ss followed by AM or PM)
%s	Seconds (00–59)
%T	Time, 24-hour (hh:mm:ss)
%W	Weekday name (Sunday–Saturday)
%w	Day of the week (0 = Sunday – 6 = Saturday)
%Y	Year, four digits
%y	Year, two digits

The datetime `2024-02-02 01:02:03` represents February 2, 2024, at 1:02 AM and 3 seconds. Try experimenting with some different formats for that `datetime`:

```
select  date_format('2024-02-02 01:02:03', '%r') as format1,
        date_format('2024-02-02 01:02:03', '%m') as format2,
        date_format('2024-02-02 01:02:03', '%M') as format3,
        date_format('2024-02-02 01:02:03', '%Y') as format4,
        date_format('2024-02-02 01:02:03', '%y') as format5,
        date_format('2024-02-02 01:02:03', '%W, %M %D at %T') as format6;
```

The result is:

```
format1      format2  format3   format4  format5  format6
-----------  -------  --------  -------  -------  -----------------------------------
01:02:03 AM    02     February   2024      24     Friday, February 2nd at 01:02:03
```

The column you aliased as `format6` shows how the format specifiers can be combined. In that format string, you added a comma and the word `at` in addition to four specifiers for the date and time.

str_to_date()

The `str_to_date()` function converts a string value to a date based on the format you provide. You use the same specifiers that you used for `date_format()`, but the two functions take opposite actions: `date_format()` converts a date to a string, while `str_to_date()` converts a string to a date.

Depending upon the format you provide, `str_to_date()` can convert a string to a `date`, a `time`, or a `datetime`:

```
select str_to_date('2024-02-02 01:02:03', '%Y-%m-%d')          as date_format,
       str_to_date('2024-02-02 01:02:03', '%Y-%m-%d %H:%i:%s') as datetime_format,
       str_to_date('01:02:03', '%H:%i:%s')                     as time_format;
```

The result is:

```
date_format  datetime_format      time_format
-----------  -------------------  -----------
2024-02-02   2024-02-02 01:02:03    01:02:03
```

The last column, time_format, can also be converted with the function of the same name. We'll look at it next.

time_format()

As its name implies, the time_format() function formats time. You can use the same specifiers as date_format() for time_format(). For example, here's how to get the current time and format it in some different ways:

```
select  time_format(curtime(), '%H:%i:%s')                                 as format1,
        time_format(curtime(), '%h:%i %p')                                 as format2,
        time_format(curtime(), '%l:%i %p')                                 as format3,
        time_format(curtime(), '%H hours, %i minutes and %s seconds') as format4,
        time_format(curtime(), '%r')                                       as format5,
        time_format(curtime(), '%T')                                       as format6;
```

Expressed in military time, the current time for me is 21:09:55, which is 9:09 PM and 55 seconds. Your results should look similar to the following:

```
format1   format2   format3  format4                              format5      format6
--------  --------  -------  -----------------------------------  -----------  --------
21:09:55  09:09 PM  9:09 PM  21 hours, 09 minutes and 55 seconds  09:09:55 PM  21:09:55
```

The column you aliased as format2 shows the hour with a leading 0 because you used the %H specifier, but the format3 column does not because you used the %h specifier. In columns 1–3, you added colon characters to the format string. In format4 you added the word hours, a comma, the word minutes, the word and, and the word seconds.

> **TRY IT YOURSELF**
>
> **8-7.** The Summer Olympics are scheduled to be held in Brisbane on July 23, 2032. Write a query to calculate how many days that is from today's date.

Mathematical Operators and Functions

MySQL provides many functions to perform calculations. There are also arithmetic operators available, like + for addition, - for subtraction, * for multiplication, / and div for division, and % and mod for modulo. You'll start reviewing some queries that use these operators, and then you'll use parentheses to control the order of operations. Afterward, you'll use mathematical functions to perform various tasks, including raising a number to a power, calculating standard deviation, and rounding and truncating numbers.

Mathematical Operators

You'll start by performing some mathematical calculations using the data from the payroll table:

```
employee    salary     deduction    bonus     tax_rate
--------  ---------  ---------  --------  --------
Max Bain   80000.00    5000.00  10000.00      0.24
Lola Joy   60000.00       0.00    800.00      0.18
Zoe Ball  110000.00    2000.00  30000.00      0.35
```

Try out some of the arithmetic operators as follows:

```
select  employee,
        salary - deduction,
        salary + bonus,
        salary * tax_rate,
        salary / 12,
        salary div 12
from    payroll;
```

In this example, you use mathematical operators to get the employee's salary minus their deductions, add their bonus to their salary, multiply their salary by their tax rate, and see their monthly salary by dividing their annual salary by 12, respectively.

The result is as follows:

```
employee salary - deduction  salary + bonus   salary * tax_rate  salary / 12  salary div 12
-------- ------------------  -------------  ------------------  -----------  -------------
Max Bain           75000.00       90000.00   9199.999570846558  6666.666667           6666
Lola Joy           60000.00       60800.00  10800.000429153442  5000.000000           5000
Zoe Ball          108000.00      140000.00   38499.99934434891  9166.666667           9166
```

Notice that in the two columns on the right, `salary / 12` and `salary div 12`, you received different results when using the / and the `div` operators. This is because `div` discards any fractional amount and / does not.

Modulo

MySQL provides two operators for modulo: the percent sign (`%`) and the `mod` operator. *Modulo* takes one number, divides it by another, and returns the remainder. Consider a table called `roulette_winning_number`:

```
winning_number
--------------
        21
         8
        13
```

You can use modulo to determine if a number is odd or even by dividing it by 2 and checking the remainder, like so:

```
select  winning_number,
        winning_number % 2
from    roulette_winning_number;
```

Anything with a remainder of 1 is an odd number. The results are as follows:

```
winning_number  winning_number % 2
--------------  ------------------
      21                1
       8                0
      13                1
```

The results show `1` for odd numbers and `0` for even numbers. In the first row, `21 % 2` evaluates to `1` because 21 divided by 2 is 10 with a remainder of 1.

NOTE *The terms* modulo *and* modulus *are often confused. Modulo is the name of a function that finds the remainder of one number being divided by another. Modulus is the number you are dividing by. For example, in the expression* `21 % 2`, `2` *is the modulus and* `%` *is the modulo operator.*

Using `mod` or `%` produces the same results. Modulo is also available as the `mod()` function. All of these queries return the same results:

```
select winning_number % 2     from roulette_winning_number;
select winning_number mod 2   from roulette_winning_number;
select mod(winning_number, 2) from roulette_winning_number;
```

Operator Precedence

When there is more than one arithmetic operator used in a mathematical expression, `*`, `/`, `div`, `%`, and `mod` are evaluated first; `+` and `-` are evaluated last. This is called *operator precedence*. The following query (which uses the `payroll` table) was written to calculate the taxes employees will pay based on their salary, bonus, and tax rate, but the query is returning the wrong tax amount:

```
select  employee,
        salary,
        bonus,
        tax_rate,
        salary + bonus * tax_rate
from    payroll;
```

The results are:

```
employee  salary     bonus     tax_rate  salary + bonus * tax_rate
--------  ---------  --------  --------  -------------------------
Max Bain   80000.00  10000.00      0.24                 82400.0000
Lola Joy   60000.00    800.00      0.18                 60144.0000
Zoe Ball  110000.00  30000.00      0.35                120500.0000
```

The column on the right should represent the amount of taxes the employees have to pay, but it seems to be too high. If Max Bain's salary is $80,000 and his bonus is $10,000, it doesn't seem reasonable that he would be required to pay $82,400 in taxes.

The query is returning the wrong value because you expected MySQL to add `salary` and `bonus` first, and then multiply the result by the `tax_rate`. Instead, MySQL multiplied `bonus` by `tax_rate` first and then added the `salary`. The multiplication happened first because multiplication has a higher operator precedence than addition.

To correct the problem, use parentheses to tell MySQL to consider `salary + bonus` as a group:

```
select  employee,
        salary,
        bonus,
        tax_rate,
        (salary + bonus) * tax_rate
from    payroll;
```

The results are:

```
employee  salary     bonus     tax_rate  salary + bonus * tax_rate
--------  ---------  --------  --------  -------------------------
Max Bain   80000.00  10000.00      0.24                 21600.0000
Lola Joy   60000.00    800.00      0.18                 10944.0000
Zoe Ball  110000.00  30000.00      0.35                 49000.0000
```

Now the query returns $21,600 for Max Bain, which is the correct value. You should use parentheses frequently when performing calculations—not only because it gives you control over the order of operations, but also because it makes your SQL easier to read and understand.

Mathematical Functions

MySQL provides many mathematical functions that can help with tasks like rounding numbers, getting the absolute value of a number, and dealing with exponents, as well as finding cosines, logarithms, and radians.

abs()

The `abs()` function gets the absolute value of a number. The absolute value of a number is always positive. For example, the absolute value of 5 is 5, and the absolute value of –5 is 5.

Say you had a contest to guess the number of jelly beans in a jar. Write a query to see whose guess was closest to the actual number, 300:

```
select  guesser,
        guess,
        300          as actual,
        300 - guess  as difference
from    jelly_bean;
```

Here you've selected the guesser's name and their guess from the jelly_bean table. You select 300 and alias the column as actual so it will appear in your results with that heading. Then you subtract the guess from 300 and alias that column as difference. The results are:

```
guesser  guess actual  difference
-------  ----- ------  ----------
Ruth       275    300          25
Henry      350    300         -50
Ike        305    300          -5
```

The difference column shows how far off the guesses were from the actual value of 300, but the results are a bit hard to interpret. When the guess was higher than the actual amount of 300, your difference column appears as a negative number. When the guess was lower than the actual amount, your difference column appears as a positive number. For your contest, you don't care whether the guess was higher or lower than 300, you only care about which guess was closest to 300.

You can use the abs() function to remove the negative numbers from the difference column:

```
select  guesser,
        guess,
        300 as actual,
        abs(300 - guess) as difference
from    jelly_bean;
```

The results are:

```
guesser  guess actual  difference
-------  ----- ------  ----------
Ruth       275    300          25
Henry      350    300          50
Ike        305    300           5
```

Now you can easily see that Ike won your contest because his value in the difference column is the smallest.

ceiling()

The ceiling() function returns the smallest whole number that is greater than or equal to the argument. If you pay $3.29 for gas, and you want to round that number up to the next whole dollar amount, you'd write the following query:

```
select ceiling(3.29);
```

The result is:

```
ceiling(3.29)
-------------
      4
```

The ceiling() function has a synonym, ceil(), that produces identical results.

floor()

The floor() function returns the largest whole number that is less than or equal to the argument. To round $3.29 down to the next lowest whole dollar amount, you'd write the following query:

```
select floor(3.29);
```

The result is:

```
floor(3.29)
-----------
     3
```

If the argument is already a whole number, then that number will be returned in both the ceiling() and floor() functions. For example, ceiling(33) and floor(33) both return 33.

pi()

The pi() function returns the value of pi, as seen at the beginning of this chapter.

degrees()

The degrees() function converts radians to degrees. You can convert pi to degrees using this query:

```
select degrees(pi());
```

The result is:

```
degrees(pi())
-------------
     180
```

You got your answer by wrapping the pi() function in the degrees() function.

radians()

The radians() function converts degrees to radians. You can convert 180 to radians using this query:

```
select radians(180);
```

Your results are:

```
radians(180)
-----------------
3.141592653589793
```

The function was sent an argument of `180` and returned a value of pi.

exp()

The `exp()` function returns the natural logarithm base number *e* raised to the power of the number you provide as an argument (2, in this example):

```
select exp(2);
```

The result is:

```
7.38905609893065
```

The function returned `7.38905609893065`, which is *e* (2.718281828459) squared.

log()

The `log()` function returns the natural logarithm of the number you provide as an argument:

```
select log(2);
```

The result is:

```
0.6931471805599453
```

MySQL also provides the `log10()` function, which returns the base-10 logarithm, and `log2()`, which returns the base-2 logarithm.

The `log()` function can accept two arguments: the base of a number, then the number itself. For example, to find the $\log_2(8)$, enter the following:

```
select log(2, 8);
```

The result is:

```
log(2, 8)
--------
    3
```

The function was sent two arguments, `2` and `8`, and returned a value of `3`.

mod()

The mod() function, as you saw earlier, is the modulo function. It takes one number, divides it by another, and returns the remainder.

```
select mod(7, 2);
```

The result is:

```
mod(7, 2)
--------
    1
```

The mod(7,2) function evaluates to 1 because 7 divided by 2 is 3 with a remainder of 1. Modulo is also available as the % operator and the mod operator.

pow()

The pow() function returns a number raised to a power. To raise 5 to the power of 3, you could write this query:

```
select pow(5, 3);
```

The result is:

```
pow(5, 3)
--------
   125
```

The pow() function has a synonym, power(), that returns identical results.

round()

The round() function, introduced earlier in the chapter, rounds decimal numbers. To round the number 9.87654321 to three digits after the decimal point, use the following query:

```
select round(9.87654321, 3);
```

The result is:

```
round(9.87654321, 3)
-------------------
        9.877
```

To round all of the fractional numbers, call round() with just one argument:

```
select round(9.87654321);
```

The result is:

```
round(9.87654321)
-------------------
        10
```

Calling round() without the optional second argument causes it to default to 0 digits after the decimal point.

truncate()

The truncate() function shortens a number to specified number of decimal places. To truncate the number 9.87654321 to three digits after the decimal point, use the following query:

```
select truncate(9.87654321, 3);
```

The result is:

```
truncate(9.87654321, 3)
-----------------------
      9.876
```

To truncate all of the fractional numbers, call truncate() with 0 as the second argument:

```
select truncate(9.87654321, 0);
```

The result is:

```
truncate(9.87654321, 0)
----------------------
         9
```

The truncate() function removes digits to convert the number to the requested number of digits after the decimal point. This differs from round(), which rounds numbers up or down before removing digits.

sin()

The sin() function returns the sine of a number given in radians. You can use this query to get the sine of 2:

```
select sin(2);
```

The result is:

```
sin(2)
------------------
0.9092974268256817
```

The function was sent an argument of 2 and returned a value of 0.9092974268256817.

cos()

The cos() function returns the cosine of a number that is given in radians. Use the following query to get the cosine of 2:

```
select cos(2);
```

The result is:

```
cos(2)
------------------
-0.4161468365471424
```

The function was sent an argument of 2 and returned a value of -0.4161468365471424.

sqrt()

The sqrt() function returns the square root of a number. You can get the square root of 16 like so:

```
sqrt(16)
--------
   4
```

The function was sent an argument of 16 and returned a value of 4.

stddev_pop()

The stddev_pop() function returns the population standard deviation of the numbers provided. *Population standard deviation* is the standard deviation when all values of a dataset are taken into consideration. For example, look at the test_score table, which contains all of your test scores:

```
score
-----
  70
  82
  97
```

Now write a query to get the population standard deviation of test scores:

```
select  stddev_pop(score)
from    test_score;
```

The result is:

```
stddev_pop(score)
-----------------
11.045361017187261
```

The std() and stddev() functions are synonyms for stddev_pop() and will produce identical results.

To get the standard deviation of a sample of values, rather than the entire population of a dataset, you can use stddev_samp() function.

tan()

The tan() function accepts an argument in radians and returns the tangent. For example, you can get the tangent of 3.8 with the following query:

```
select tan(3.8);
```

The result is:

```
0.7735560905031258
```

The function was sent an argument of 3.8 and returned a value of 0.7735560905031258.

> **TRY IT YOURSELF**
>
> **8-8.** The moon is 252,088 miles from Earth. Write a query to calculate how many kilometers the moon is from Earth. Round the number to the nearest kilometer. To convert miles to kilometers, multiply the miles by 1.60934.

Other Handy Functions

Other useful functions include cast(), coalesce(), distinct(), database(), if(), and version().

cast()

The cast() function converts a value from one data type to a different data type. To call the cast() function, pass a value into cast() as the first argument, follow it with the as keyword, and then specify the data type you want to convert it to.

For example, select the datetime column order_datetime from the table called online_order:

```
select  order_datetime
from    online_order;
```

Your results show the following datetime values:

```
order_datetime
-------------------
2024-12-08 11:39:09
2024-12-10 10:11:14
```

You can select those values without their time portion by casting from a datetime data type to a date data type, like so:

```
select  cast(order_datetime as date)
from    online_order;
```

Your results are:

```
cast(order_datetime as date)
----------------------------
         2024-12-08
         2024-12-10
```

The date part of the datetime now appears as a date value.

coalesce()

The coalesce() function returns the first non-null value in a list. You could specify null values followed by a non-null value, and coalesce() would return the non-null value:

```
select coalesce(null, null, 42);
```

The result is:

```
coalesce(null, null, 42)
------------------------
            42
```

The coalesce() function is also useful when you want to display a value in your result instead of null. For example, in the candidate table used in the following query, the employer column will sometimes store the candidate's employer name, and other times that column will be null. In order to display the text Between Jobs instead of null, you'd enter the following:

```
select employee_name,
       coalesce(employer, 'Between Jobs')
from   candidate;
```

The results are:

```
employee_name  employer
-------------  ------------
Jim Miller     Acme Corp
Laura Garcia   Globex
Jacob Davis    Between Jobs
```

The query now displays Between Jobs rather than null for Jacob Davis, which is more informative, especially for nontechnical users who may not understand what null means.

distinct()

When you have duplicate values, you can use the `distinct()` function to display each value only once. For example, if you want to know which countries your customers are from, you could query the `customer` table like so:

```
select country
from   customer;
```

The result is:

```
country
-------
India
USA
USA
USA
India
Peru
```

The query is returning the `country` column value for every row in the `customer` table. You can use the `distinct()` function to see each country in your result set just once:

```
select distinct(country)
from   customer;
```

Now the result is:

```
country
-------
India
USA
Peru
```

The `distinct()` function is also available as an operator. To use it, remove the parentheses like so:

```
select distinct country
from   customer;
```

The result set is identical:

```
country
-------
India
USA
Peru
```

The distinct() function is especially useful when combined with the count() function to find how many unique values you have. Here you write a query to count the number of distinct countries in your table:

```
select count(distinct country)
from   customer;
```

The result is:

```
count(distinct country)
-----------------------
          3
```

You identified the distinct countries using the distinct() function and wrapped them in the count() function to get a count of them.

database()

The database() function tells you which database you're currently using. As you saw in Chapter 2, the use command lets you select which database you want to use. Throughout your day, you might move between different databases and forget your current database. You can call the database() function like so:

```
use airport;

select database();
```

The result is:

```
database()
----------
 airport
```

If you're not in the database you thought you were and you tried to query a table, MySQL would give an error saying the table doesn't exist. Calling database() is a quick way to check.

if()

The if() function returns a different value depending upon whether a condition is true or false. The if() function accepts three arguments: the condition you want to test, the value to return if the condition is true, and the value to return if the condition is false.

Let's write a query that lists students and whether they passed or failed an exam. The test_result table contains the following data:

```
student_name  grade
------------  -----
Lisa          98
Bart          41
Nelson        11
```

Your query to check if each student passed the exam should look similar to the following:

```
select  student_name,
        if(grade > 59, 'pass', 'fail')
from    test_result;
```

The condition you're testing is whether the student's grade is greater than 59. If so, you return the text pass. If not, you return the text fail. The results are:

```
student_name  if(grade > 59, 'pass', 'fail')
------------  ------------------------------
Lisa                       pass
Bart                       fail
Nelson                     fail
```

NOTE *There is also an* if *statement that is described in Chapter 11 that you'll use when you create your own functions and procedures. This statement is different from the* if() *function described here.*

MySQL also has a case operator that lets you perform more sophisticated logic than the if() function. The case operator lets you test more than one condition and returns the result for the first condition that is met. In the following query, you select the student name and add a comment to the student based on their grade:

```
select  student_name,
case
  when grade < 30 then 'Please retake this exam'
  when grade < 60 then 'Better luck next time'
  else 'Good job'
end
from test_result;
```

The case operator uses a matching end keyword that marks the end of the case statement.

For any students who received a grade less than 30, the case statement returns Please retake this exam and then control is passed to the end statement.

Students who received a grade of 30 or more aren't handled by the first when condition of the case statement, so control drops to the next line.

If a student received a grade of 30 or higher but less than 60, Better luck next time is returned and control passes to the end statement.

If a student's grade didn't match either of the when conditions, meaning the student scored higher than 60, control drops to the else keyword, where Good job is returned. You use an else clause to capture any student grades that aren't handled by the first two conditions. The results are:

```
student_name  case when grade < 30 then 'Please...
------------  -----------------------------------
Lisa          Good job
```

```
Bart          Better luck next time
Nelson        Please retake this exam
```

Unlike the if() function—which returns a result if a condition is true or false—case lets you check several conditions and returns a result based on the first condition that is met.

version()

The version() function returns the version of MySQL you are using:

```
select version();
```

The result is:

```
version
-------
8.0.27
```

The version of MySQL installed on my server is 8.0.27. Yours may be different.

TRY IT YOURSELF

8-9. In the electricity database, the electrician table contains the following data:

```
electrician     years_experience
-------------   ----------------
Zach Zap               1
Wanda Wiring           6
Larry Light           21
```

In this company, an electrician with less than 5 years' experience gets the Journeyman title; an electrician with between 5 and 10 years' experience has the Apprentice title; and the Master Electrician title is given to those with 10 years' experience or more.

The following query was written to display each electrician's name and title, but it is returning the wrong results:

```
select  name,
case
  when years_experience < 10 then 'Apprentice'
  when years_experience < 5 then 'Journeyman'
  else 'Master Electrician'
end
from    electrician;
```

The results are:

```
name            case when years_e...
--------------  ------------------
Zach Zap        Apprentice
Wanda Wiring    Apprentice
Larry Light     Master Electrician
```

Zach Zap should be appearing as a Journeyman, not an Apprentice. What is wrong with the query? Modify the query to return the correct names and titles.

Summary

In this chapter, you looked at how to call MySQL built-in functions and pass values, known as arguments, to those functions. You explored the most useful functions and saw how to locate the more obscure ones when necessary. In the next chapter, you'll look at how to insert, update, and delete data from a MySQL database.

9

INSERTING, UPDATING, AND DELETING DATA

In this chapter, you'll learn to insert, update, and delete data from tables. You'll practice ways to insert data from one table to another, use queries to update or delete data from a table, and create a table that automatically increments a numeric value into a column as you insert rows.

Inserting Data

So far, you've been querying data from tables. But how did the data get into the tables in the first place? Typically, you insert data using the `insert` statement.

Adding rows to a table with the `insert` statement is known as *populating* a table. You specify the name of the table, the names of the columns you want to insert values into, and the values you want to insert.

Here you insert a row of data into the arena table, which contains information about various arena names, locations, and capacities:

```
❶ insert into arena
      (
    ❷ arena_id,
      arena_name,
      location,
      seating_capacity
      )
❸ values
      (
      1,
    ❹ 'Madison Square Garden',
      'New York',
      20000
      );
```

First, you specify that you want to insert a row into the arena table ❶, and that your data will go into the arena_id, arena_name, location, and seating_capacity columns ❷. You then list the values you want to insert under the values keyword in the same order in which you listed the columns ❸. You surround, or wrap, the values Madison Square Garden and New York in quotes because they are character strings ❹.

When you run this insert statement, MySQL returns the message 1 row(s) affected to let you know that one row was inserted into the table.

You can then query your arena table to confirm the new row looks as you intended:

```
select * from arena;
```

The result is:

```
arena_id  arena_name             location  seating_capacity
--------  ---------------------  --------  ----------------
    1     Madison Square Garden  New York        20000
```

The row was inserted, and the columns and their values appear as you expected.

Inserting Null Values

When you want to insert a null value into a column, you have two options. First, you can list the column name and use the null keyword as the value to insert. For example, if you want to add a row to the arena table for the Dean Smith Center but don't know its seating capacity, you can write an insert statement like this:

```
insert into arena
    (
    arena_id,
```

```
    arena_name,
    location,
    seating_capacity
    )
values
    (
    2,
    'Dean Smith Center',
    'North Carolina',
    null
    );
```

The second option is to omit the column name entirely. As an alternative to the preceding insert statement, you can omit the seating_capacity column from your list of columns and provide no value for it in your list of values:

```
insert into arena
    (
    arena_id,
    arena_name,
    location
    )
values
    (
    2,
    'Dean Smith Center',
    'North Carolina'
    );
```

Since you didn't insert a value into the seating_capacity column, MySQL will set it to null by default. You can see the row that was inserted using this query:

```
select  *
from    arena
where   arena_id = 2;
```

The result is:

```
arena_id  arena_name         location        seating_capacity
--------  -----------------  --------        ----------------
   2      Dean Smith Center  North Carolina        null
```

The seating_capacity column will be set to null regardless of which approach you take.

If the seating_capacity column had been defined as not null when you created the table, you wouldn't be allowed to insert a null value using either approach (see Chapter 2).

TRY IT YOURSELF

9-1. Create a database called `food`. In the database, create a table called `favorite_meal` that has two columns. The `meal` column should be defined as `varchar(50)`, and the `price` column should be defined as `numeric(5,2)`. Then insert the following data into the table:

```
meal          price
------------  -------
Pizza          7.22
Cheeseburger   8.41
Salad          9.57
```

Run the query **`select * from favorite_meal;`** to see your new rows in the table.

9-2. Create a database called `education`. In the database, create a table called `college` that has three columns. The `college_name` column should be defined as `varchar(100)`, the `location` column should be defined as `varchar(50)`, and the `undergrad_enrollment` column should be defined as an `int`. Insert the following data into the table:

```
college_name                          location        undergrad_enrollment
------------------------------------  --------        --------------------
Princeton University                  New Jersey              4773
Massachusetts Institute of Technology Massachusetts           4361
Oxford University                     Oxford                 11955
```

Run the query **`select * from college;`** to see your new rows in the table.

Inserting Multiple Rows at Once

When you want to insert multiple rows, you can either insert one row at a time or insert them as a group. Let's start with the first approach. Here's how you insert three arenas into the arena table using individual insert statements:

```
insert into arena (arena_id, arena_name, location, seating_capacity)
values (3, 'Philippine Arena', 'Bocaue', 55000);

insert into arena (arena_id, arena_name, location, seating_capacity)
values (4, 'Sportpaleis', 'Antwerp', 23359);

insert into arena (arena_id, arena_name, location, seating_capacity)
values (5, 'Bell Centre', 'Montreal', 22114);
```

You could achieve the same results by combining all three rows into one insert statement:

```
insert into arena (arena_id, arena_name, location, seating_capacity)
values (3, 'Philippine Arena', 'Bocaue', 55000),
       (4, 'Sportpaleis', 'Antwerp', 23359),
       (5, 'Bell Centre', 'Montreal', 22114);
```

To insert multiple rows at once, surround each row's values with parentheses and use a comma between each set of values. MySQL will insert all three rows into the table and give you the message `3 row(s) affected` to let you know that all three rows were inserted.

NOTE *The format of these SQL statements is a little different from the previous examples you have seen. The list of column names (`arena_id`, `arena_name`, `location`, and `seating_capacity`) is all on one line. You might prefer to put your column names on one line like this to save space, or you might find it more readable to list each column on its own line, as in the earlier examples. The choice is yours.*

Inserting Without Listing Column Names

You can also insert data into a table without specifying the column names. Since you're inserting four values and the `arena` table only has four columns, you could replace the `insert` statement that lists the column names with one that does not:

```
insert into arena
values (6, 'Staples Center', 'Los Angeles', 19060);
```

MySQL is able to determine which columns to insert the values into because you've provided the data in the same order as the columns in your table.

Although omitting the column names saves you some typing, it's best practice to list them. At some point in the future, you might add a fifth column to the `arena` table. If you don't list your columns, making that change would break your `insert` statements because you'd be trying to insert four values into a table with five columns.

Inserting Sequences of Numbers

You might want to insert sequential numbers into a table column, such as in the `arena` table where the first row of the `arena_id` column should have the value `1`, the next row of the `arena_id` column should have the value `2`, the next row should have a value of `3`, and so on. MySQL provides an easy way to do that by letting you define a column with the `auto_increment` attribute. The `auto_increment` attribute is particularly useful for a primary key column—that is, the column that uniquely identifies the rows in a table.

Let's look at how it works. Select everything from the arena table you've created thus far:

```
select * from arena;
```

The results are:

```
arena_id  arena_name             location        seating_capacity
--------  ---------------------  --------------  ----------------
   1      Madison Square Garden  New York             20000
   2      Dean Smith Center      North Carolina        null
   3      Philippine Arena       Bocaue               55000
   4      Sportpaleis            Antwerp              23359
   5      Bell Centre            Montreal             22114
   6      Staples Center         Los Angeles          19060
```

You can see that each arena has its own arena_id that is one larger than the value for the arena that was inserted before it.

When you inserted the values in the arena_id column, you found the highest arena_id already in the table and added 1 to it when inserting the next row. For example, when you inserted the row for the Staples Center, you hardcoded the arena_id as 6 because the previous arena_id was 5:

```
insert into arena (arena_id, arena_name, location, seating_capacity)
values (6, 'Staples Center', 'Los Angeles', 19060);
```

This approach won't work very well in a real production database where many new rows are being created quickly. A better approach is to have MySQL manage that work for you by defining the arena_id column with auto_increment when you create the table. Let's try it.

Drop the arena table and re-create it using auto_increment for the arena_id column:

```
drop table arena;

create table arena (
    arena_id          int           primary key       auto_increment,
    arena_name        varchar(100),
    location          varchar(100),
    seating_capacity  int
);
```

Now when you insert rows into the table, you won't have to deal with inserting data into the arena_id column. You can insert data into the other columns and MySQL will automatically increment the arena_id column for you with each new row that you insert. Your insert statements should look like this:

```
insert into arena (arena_name, location, seating_capacity)
values ('Madison Square Garden', 'New York', 20000);

insert into arena (arena_name, location, seating_capacity)
```

```
values ('Dean Smith Center', 'North Carolina', null);

insert into arena (arena_name, location, seating_capacity)
values ('Philippine Arena', 'Bocaue', 55000);

insert into arena (arena_name, location, seating_capacity)
values ('Sportpaleis', 'Antwerp', 23359);

insert into arena (arena_name, location, seating_capacity)
values ('Bell Centre', 'Montreal', 22114);

insert into arena (arena_name, location, seating_capacity)
values ('Staples Center', 'Los Angeles', 19060);
```

You didn't list `arena_id` as one of the columns in your list of columns, nor did you provide a value for `arena_id` in your list of values. Take a look at the rows in the table after MySQL runs your `insert` statements:

```
select * from arena;
```

The results are:

```
arena_id  arena_name             location        seating_capacity
--------  ---------------------  --------------  ----------------
   1      Madison Square Garden  New York              20000
   2      Dean Smith Center      North Carolina         null
   3      Philippine Arena       Bocaue                55000
   4      Sportpaleis            Antwerp               23359
   5      Bell Centre            Montreal              22114
   6      Staples Center         Los Angeles           19060
```

As you can see, MySQL automatically incremented the values for the `arena_id` column.

Only one column per table can be defined with `auto_increment`, and it has to be the primary key column (or a column that is part of the primary key).

When inserting a value into a column defined with `auto_increment`, MySQL will always insert a higher number, but there can be gaps between the numbers. For example, you could end up with `arena_id` 22, 23, and then 29 in your table. The reasons for this have to do with the storage engine your database is using, how your MySQL server is configured, and other factors that are beyond the scope of this book, so just know that a column defined with `auto_increment` will always result in an ascending list of numbers.

Inserting Data Using a Query

You can insert data into a table based on values returned from a query. For example, say the `large_building` table has data you want to add to your `arena` table. The `large_building` table was created with these data types:

```
create table large_building
       (
       building_type      varchar(50),
```

```
    building_name      varchar(100),
    building_location  varchar(100),
    building_capacity  int,
    active_flag        bool
);
```

It contains this data:

```
building_type  building_name      building_location  building_capacity  active_flag
-------------  -----------------  -----------------  -----------------  -----------
Hotel          Wanda Inn          Cape Cod                  125              1
Arena          Yamada Green Dome  Japan                   20000              1
Arena          Oracle Arena       Oakland                 19596              1
```

For your purposes, you don't care about the first row in the table, because `Wanda Inn` is a hotel, not an arena. You can write a query to return the arena data from the other rows in the `large_building` table like so:

```
select   building_name,
         building_location,
         building_capacity
from     large_building
where    building_type = 'Arena'
and      active_flag is true;
```

The results are:

```
building_name      building_location  building_capacity
-----------------  -----------------  -----------------
Yamada Green Dome  Japan                   20000
Oracle Arena       Oakland                 19596
```

You can then use that query as the basis for an insert statement to insert these rows into the arena table:

```
insert into arena (
         arena_name,
         location,
         seating_capacity
)
select   building_name,
         building_location,
         building_capacity
from     large_building
where    building_type = 'Arena'
and      active_flag is true;
```

MySQL inserts the two rows that were returned from your query into the arena table. You can query the arena table to see the new rows:

```
select * from arena;
```

Here are the results with the new rows included:

```
arena_id  arena_name             location        seating_capacity
--------  ---------------------  --------------  ----------------
    1     Madison Square Garden  New York              20000
    2     Dean Smith Center      North Carolina         null
    3     Philippine Arena       Bocaue                55000
    4     Sportpaleis            Antwerp               23359
    5     Bell Centre            Montreal              22114
    6     Staples Center         Los Angeles           19060
    7     Yamada Green Dome      Japan                 20000
    8     Oracle Arena           Oakland               19596
```

The insert statement added arenas 7 and 8 to the existing data in the arena table.

Using a Query to Create and Populate a New Table

The create table as syntax allows you to create and populate a table in one step. Here you create a new table called new_arena and insert rows into it at the same time:

```
create table new_arena as
select  building_name,
        building_location,
        building_capacity
from    large_building
where   building_type = 'Arena'
and     active_flag is true;
```

NOTE *The as keyword is optional.*

This statement creates a table called new_arena based on the results of the preceding large_building query. Now query the new table:

```
select * from new_arena;
```

The results are:

```
building_name      building_location  building_capacity
-----------------  -----------------  -----------------
Yamada Green Dome  Japan                    20000
Oracle Arena       Oakland                  19596
```

The new_arena table is created with the same column names and data types as the large_building table. You can confirm the data types by describing the table with the desc keyword:

```
desc new_arena;
```

The results are:

```
Field              Type          Null  Key  Default  Extra
-----------------  ------------  ----  ---  -------  -----
building_name      varchar(100)  YES         null
building_location  varchar(100)  YES         null
building_capacity  int           YES         null
```

You can also use `create table` to make a copy of a table. For example, you might save the current state of the arena table by making a copy of it and calling the new table arena_ with the current date appended to it, like so:

```
create table arena_20241125 as
select * from arena;
```

Before you add or remove columns from the arena table, you might want to ensure you have your original data saved in a second table first. This is useful when you're about to make major changes to a table, but it may not be practical to make a copy of a very large table.

Updating Data

Once you have data in your tables, you'll likely want to make changes to it over time. MySQL's `update` statement allows you to modify existing data.

Arenas are notorious for having their names changed, and the arenas in your table are no exception. Here you change the `arena_name` value for `arena_id 6` from `Staples Center` to `Crypto.com Arena` using the `update` statement:

```
update  arena
set     arena_name = 'Crypto.com Arena'
where   arena_id = 6;
```

First, you use the set keyword to set column values in the table. Here you are setting the `arena_name` column's value to `Crypto.com Arena`.

Next, you specify which row(s) you want updated in the `where` clause. In this case, you chose to update the row based on the `arena_id` column with a value of `6`, but you could have updated that same row based on another column. For example, you can update the row based on the `arena_name` column instead:

```
update  arena
set     arena_name = 'Crypto.com Arena'
where   arena_name = 'Staples Center';
```

Or, since you have only one arena in Los Angeles listed, you can update the row using the `location` column:

```
update  arena
set     arena_name = 'Crypto.com Arena'
where   location = 'Los Angeles';
```

It's important that you craft your where clauses carefully because any rows that match the criteria specified there will be updated. For example, if there are five arenas with a location of Los Angeles, this update statement will rename all five to Crypto.com Arena, whether or not that's what you intended.

It's usually best to update rows based on a primary key column. When you created the arena table, you defined the arena_id column as the primary key of the table. That means there will only be one row in the table for an arena_id of 6, so if you use the syntax where arena_id = 6, you can be confident you're updating only that row.

Using a primary key in your where clause is also best practice because primary key columns are indexed. Indexed columns are typically faster at finding rows in the table than unindexed columns.

Updating Multiple Rows

To update multiple rows, you can use a where clause that matches more than one row. Here you update the seating capacity of all arenas with an arena_id greater than 3:

```
update  arena
set     seating_capacity = 20000
where   arena_id > 3;
```

MySQL updates arenas 4, 5, and 6 to have seating_capacity values of 20,000.

If you remove your where clause entirely, all rows in your table will be updated:

```
update  arena
set     seating_capacity = 15000;
```

If you select * from arena now, you can see that all arenas have a seating capacity of 15,000:

```
arena_id  arena_name              location        seating_capacity
--------  ----------------------  --------------  ----------------
    1     Madison Square Garden   New York             15000
    2     Dean Smith Center       North Carolina       15000
    3     Philippine Arena        Bocaue               15000
    4     Sportpaleis             Antwerp              15000
    5     Bell Centre             Montreal             15000
    6     Crypto.com Arena        Los Angeles          15000
```

In this example, it's apparent that you forgot to use a where clause to limit the number of rows to update.

Updating Multiple Columns

You can update more than one column with one update statement by separating the column names with a comma:

```
update  arena
set     arena_name = 'Crypto.com Arena',
```

```
        seating_capacity = 19100
where   arena_id = 6;
```

Here, you've updated both the arena_name and the seating_capacity column values for the row that has an arena_id of 6.

> **TRY IT YOURSELF**
>
> **9-3.** In the food database, update all prices in the favorite_meal table so that they're raised by 20 percent.

Deleting Data

To remove data from your tables, you use the delete statement. You can delete one row at a time, multiple rows, or all rows with one delete statement. You use the where clause to specify which rows you want to delete. Here, you delete the row with an arena_id of 2:

```
delete from arena
where arena_id = 2;
```

After you run this delete statement, select the remaining rows from the table like so:

```
select * from arena;
```

The result is:

```
arena_id  arena_name             location        seating_capacity
--------  ---------------------  --------------  ----------------
    1     Madison Square Garden  New York            15000
    3     Philippine Arena       Bocaue              15000
    4     Sportpaleis            Antwerp             15000
    5     Bell Centre            Montreal            15000
    6     Crypto.com Arena       Los Angeles         15000
```

You can see that the row containing the arena_id of 2 has been deleted.

In Chapter 7, you learned about using like for simple pattern matches. You can do that here to delete all arenas that have the word Arena in their name:

```
delete from arena
where arena_name like '%Arena%';
```

Select the remaining rows from the table:

```
select * from arena;
```

The result is:

```
arena_id  arena_name             location       seating_capacity
--------  ---------------------  -------------  ----------------
    1     Madison Square Garden  New York            15000
    4     Sportpaleis            Antwerp             15000
    5     Bell Centre            Montreal            15000
```

The two rows containing `Philippine Arena` and `Crypto.com Arena` are no longer in the table.

If you write a `delete` statement and the `where` clause doesn't match any rows, no rows will be deleted:

```
delete from arena
where arena_id = 459237;
```

This statement won't delete any rows because there aren't any with an `arena_id` of `459237`. MySQL won't produce an error message, but it will tell you `0 row(s) affected`.

To delete all rows from the table, you can use a `delete` statement without a `where` clause:

```
delete from arena;
```

This statement removes all rows from the table.

NOTE *As with* `update` *statements, you need to take care with your* `where` *clauses when writing* `delete` *statements. MySQL will delete all the rows identified in the* `where` *clause, so be sure that it's correct.*

TRY IT YOURSELF

9-4. Due to a mozzarella shortage, you need to remove `Pizza` from the `favorite_meal` table in the `food` database. Write a `delete` statement that accomplishes this.

Truncating and Dropping a Table

Truncating a table removes all the rows but keeps the table intact. It has the same effect as using `delete` without a `where` clause, but it's typically faster.

You can truncate a table using the `truncate table` command, like so:

```
truncate table arena;
```

Once the statement runs, the table will still exist but there will be no rows in it.

If you want to remove both the table and all of its data, you can use the `drop table` command:

```
drop table arena;
```

If you try to select from the `arena` table now, MySQL will display a message saying the table doesn't exist.

Summary

In this chapter you looked at inserting, updating, and deleting data from a table. You saw how to insert null values and quickly create or delete entire tables. In the next chapter, you'll learn the benefits of using table-like structures called *views*.

PART III

DATABASE OBJECTS

In Part III, you'll create database objects like views, functions, procedures, triggers, and events. These objects will be stored on your MySQL server so you can call them whenever you need them.

In Chapter 10, you'll learn how to create views that let you access the results of a query as a table-like structure.

In Chapter 11, you'll create your own functions and procedures to perform tasks like getting and updating the population of states.

In Chapter 12, you'll create your own triggers that automatically take an action you define when rows are inserted, updated, or deleted from a table.

In Chapter 13, you'll create your own MySQL events to manage scheduled tasks.

In these chapters, you'll use the following naming conventions for different types of objects:

`beer`	A table that contains data about beer.
`v_beer`	A view that contains data about beer.
`f_get_ipa()`	A function that gets a list of India pale ales.
`p_get_pilsner()`	A procedure that gets a list of pilsner beers.

tr_beer_ad	A trigger that automatically takes an action after some rows in the beer table are deleted. I use the tr_ prefix for triggers so that they won't be confused with tables, which also start with the letter *t*. The suffix _ad stands for *after delete*. _bd stands for *before delete*. _bu and _au stand for *before* and *after update*, respectively. _bi and _ai stand for *before* and *after insert*, respectively. You'll learn what those suffixes mean in Chapter 12.
e_load_beer	A scheduled event to load new beer data into the beer table.

In previous chapters, you've named tables descriptively so that other programmers can quickly understand the nature of the data that the table is storing. For database objects other than tables, you'll continue using that approach and also prefix the name of the object with a short description of its type (as in v_ for *view*); occasionally, you'll add a suffix as well (as in _ad for *after delete*).

While these naming conventions aren't law, consider using them, as they help you quickly understand a database object's purpose.

10

CREATING VIEWS

In this chapter, you'll learn how to create and use views. *Views* are virtual tables based on the output of a query you write to customize the display of your result set. Each time you select from a view, MySQL reruns the query that you defined the view with, returning the latest results as a table-like structure with rows and columns.

Views are useful in situations where you want to simplify a complex query or hide sensitive or irrelevant data.

Creating a New View

You create a view using the `create view` syntax. Let's look at an example with the following `course` table:

```
course_name                               course_level
----------------------------------------  ------------
Introduction to Python                    beginner
```

```
Introduction to HTML                      beginner
React Full-Stack Web Development          advanced
Object-Oriented Design Patterns in Java   advanced
Practical Linux Administration            advanced
Learn JavaScript                          beginner
Advanced Hardware Security                advanced
```

Here you create a view named v_course_beginner that selects all columns with a course_level of beginner from the course table:

```
create view v_course_beginner as
select *
from   course
where  level = 'beginner';
```

Running this statement creates the view and saves it in your MySQL database. Now you can query the v_course_beginner view at any time, like so:

```
select * from v_course_beginner;
```

The results are:

```
course_name             course_level
----------------------  ------------
Introduction to Python  beginner
Introduction to HTML    beginner
Learn JavaScript        beginner
```

Since you defined the view by selecting * (the wildcard character) from the course table, it has the same column names as the table.

The v_course_beginner view should be used by beginner students, so you selected only courses from the table with a course_level of beginner, hiding the advanced courses.

Now create a second view for advanced students that includes just advanced courses:

```
create view v_course_advanced as
select *
from   courses
where  level = 'advanced';
```

Selecting from the v_course_advanced view displays the advanced courses:

```
select * from v_course_advanced;
```

The results are:

```
course_name                               course_level
----------------------------------------  ------------
React Full-Stack Web Development          advanced
Object-Oriented Design Patterns in Java   advanced
Practical Linux Administration            advanced
Advanced Hardware Security                advanced
```

When you defined the `v_course_advanced` view, you provided MySQL with a query that selects data from the `course` table. MySQL runs this query each time the view is used, meaning that the view is always up to date with the latest rows from the `course` table. In this example, any new advanced courses added to the `course` table will be shown each time you select from the `v_course_advanced` view.

This approach allows you to maintain your courses in the `course` table and provide different views of the data to beginner and advanced students.

Using Views to Hide Column Values

In the `course` table example, you created views that displayed certain rows from the table and hid others. You can also create views that display different *columns*.

Let's look at an example of using views to hide sensitive column data. You have two tables, `company` and `complaint`, that help track complaints for local companies.

The `company` table is as follows:

```
company_id  company_name          owner          owner_phone_number
----------  --------------------  -------------  ------------------
1           Cattywampus Cellular  Sam Shady        784-785-1245
2           Wooden Nickel Bank    Oscar Opossum    719-997-4545
3           Pitiful Pawn Shop     Frank Fishy      917-185-7911
```

And here's the `complaint` table:

```
complaint_id  company_id  complaint_desc
------------  ----------  ------------------------------
1                  1      Phone doesn't work
2                  1      Wi-Fi is on the blink
3                  1      Customer service is bad
4                  2      Bank closes too early
5                  3      My iguana died
6                  3      Police confiscated my purchase
```

You'll start by writing a query to select information about each company and a count of its received complaints:

```
select   a.company_name,
         a.owner,
         a.owner_phone_number,
         count(*)
from     company a
join     complaint b
on       a.company_id = b.company_id
group by a.company_name,
         a.owner,
         a.owner_phone_number;
```

The results are:

```
company_name         owner          owner_phone_number  count(*)
-------------------- -------------- ------------------  --------
Cattywampus Cellular Sam Shady         784-785-1245          3
Wooden Nickel Bank   Oscar Opossum     719-997-4545          1
Pitiful Pawn Shop    Frank Fishy       917-185-7911          2
```

To display the results of this query in a view called `v_complaint`, simply add the `create view` syntax as the first line of the original query:

```
create view v_complaint as
select    a.company_name,
          a.owner,
          a.owner_phone_number,
          count(*)
from      company a
join      complaint b
on        a.company_id = b.company_id
group by a.company_name,
          a.owner,
          a.owner_phone_number;
```

Now, the next time you want to get a list of companies with a count of complaints, you can simply type `select * from v_complaint` instead of rewriting the entire query.

Next, you'll create another view that hides the owner information. You'll name the view `v_complaint_public`, and you'll let all users of your database access the view. This view will show the company name and number of complaints, but not the owner's name or phone number:

```
create view v_complaint_public as
select    a.company_name,
          count(*)
from      company a
join      complaint b
on        a.company_id = b.company_id
group by a.company_name;
```

You can query the view like so:

```
select * from v_complaint_public;
```

The results are:

```
company_name         count(*)
-------------------- --------
Cattywampus Cellular    3
Wooden Nickel Bank      1
Pitiful Pawn Shop       2
```

This is an example of using a view to hide data stored in columns. While the owners' contact information is in your database, you are withholding it by not selecting those columns in your v_complaint_public view.

Once you've created your views, you can use them as if they were tables. For example, you can join views to tables, join views to other views, and use views in subqueries.

Inserting, Updating, and Deleting from Views

In Chapter 9 you learned how to insert, update, and delete rows from tables. In some cases, it's also possible to modify rows using a view. For example, the v_course_beginner view is based on the course table. You can update that view using the following update statement:

```
update  v_course_beginner
set     course_name = 'Introduction to Python 3.1'
where   course_name = 'Introduction to Python';
```

This update statement updates the course_name column in the v_course_beginner view's underlying course table. MySQL is able to perform the update because the view and the table are so similar; for every row in the v_course_beginner view, there is one row in the course table.

Now, try to update the v_complaint view with a similar query:

```
update  v_complaint
set     owner_phone_number = '578-982-1277'
where   owner = 'Sam Shady';
```

You receive the following error message:

```
Error Code: 1288. The target table v_complaint of the UPDATE is not updatable
```

MySQL doesn't allow you to update the v_complaint view, because it was created using multiple tables and the count() aggregate function. It's a more complex view than the v_course_beginner view. The rules about which views allow rows to be updated, inserted, or deleted are fairly complicated. For this reason, I recommend changing data directly from tables and avoiding using views for this purpose.

Dropping a View

To remove a view, use the drop view command:

```
drop view v_course_advanced;
```

While the view is removed from the database, the underlying table still exists.

TRY IT YOURSELF

The corporate database contains the following employee table:

```
employee_name   department  position    home_address       date_of_birth
--------------  ----------  ----------  -----------------  -------------
Sidney Crumple  accounting  Accountant  123 Credit Road       1997-01-04
Al Ledger       accounting  Bookkeeper  2 Revenue Street      2002-11-22
Bean Counter    accounting  Manager     8 Bond Street         1996-04-29
Lois Crumple    accounting  Accountant  123 Debit Lane        2000-08-27
Lola Hardsell   sales       Sales Rep   66 Hawker Street      2000-07-09
Bob Closer      sales       Sales Rep   73 Peddler Way        1999-02-16
```

10-1. Create a view called v_employee_accounting that has all employees in accounting.

10-2. Create a view called v_employee_sales that has all employees in sales.

10-3. Create a view called v_employee_private that has all employees in all departments. Hide the home_address and date_of_birth columns. Create three queries that select * from each view. Do they return what you expect?

Indexes and Views

You can't add indexes to views to speed up your queries, but MySQL can use any indexes on the underlying tables. For example, the following query

```
select  *
from    v_complaint
where   company_name like 'Cattywampus%';
```

can take advantage of an index on the company_name column of the company table, since the v_complaint view is built on the company table.

Summary

In this chapter, you saw how to use views to provide a custom representation of your data. In the next chapter, you'll learn how to write functions and procedures and add logic to them to perform certain tasks based on your data values.

11

CREATING FUNCTIONS AND PROCEDURES

In Chapter 8, you learned how to call built-in MySQL functions; in this chapter, you'll write your own. You'll also learn to write procedures and explore the key differences between the two.

You'll add logic to your functions and procedures using `if` statements, loops, cursors, and `case` statements to perform different tasks based on the value of your data. Lastly, you'll practice accepting values in your functions and procedures and returning values.

Functions vs. Procedures

Functions and procedures are programs you can call by name. Because they're saved in your MySQL database, they are sometimes called *stored* functions and procedures. Collectively, they are referred to as *stored routines* or *stored programs*. When you write a complex SQL statement or a group of statements with several steps, you should save it as a function or procedure so you can easily call it by name later.

The main difference between a function and a procedure is that a function gets called from a SQL statement and always returns one value. A procedure, on the other hand, gets called explicitly using a `call` statement. Procedures also pass values back to the caller differently than functions. (Note that the caller may be a person using a tool like MySQL Workbench, a program written in a programming language like Python or PHP, or another MySQL procedure.) While procedures may return no values, one value, or many values, a function accepts arguments, performs some task, and returns a single value. For example, you might find the population of New York by calling the `f_get_state_population()` function from a `select` statement, passing in the state name as an argument to the function:

```
select f_get_state_population('New York');
```

You pass an argument to a function by putting it between the parentheses. To pass more than one argument, separate them by commas. The function accepts the argument, does some processing that you defined when you created the function, and returns a value:

```
f_get_state_population('New York')
----------------------------------
                19299981
```

The `f_get_state_population()` function took the text `New York` as an argument, did a lookup in your database to find the population, and returned `19299981`.

NOTE *Consider starting your custom functions with f_ to make their role clear, as I've done here.*

You can also call functions in the `where` clause of SQL statements, such as the following example that returns every `state_population` greater than New York's:

```
select  *
from    state_population
where   population > f_get_state_population('New York');
```

Here, you called the function `f_get_state_population()` with an argument of `New York`. The function returned the value `19299981`, which caused your query to evaluate to the following:

```
select  *
from    state_population
where   population > 19299981;
```

Your query returned data from the state table for states with a population greater than 19,299,981:

```
state       population
----------  ----------
California  39613493
```

```
Texas        29730311
Florida      21944577
```

Procedures, on the other hand, are not called from SQL queries, but instead via the `call` statement. You pass in any arguments the procedure has been designed to accept, the procedure performs the tasks you defined, and then control returns to the caller.

For example, you call a procedure named `p_set_state_population()` and pass it an argument of `New York` like so:

```
call p_set_state_population('New York');
```

You'll see how to create the `p_set_state_population()` procedure and define its tasks in Listing 11-2. For now, just know that this is the syntax for calling a procedure.

NOTE *Consider starting procedures with* `p_` *to make their role clear.*

Procedures are often used to execute business logic by updating, inserting, and deleting records in tables, and they can also be used to display a dataset from the database. Functions are used for smaller tasks, like getting one piece of data from the database or formatting a value. Sometimes you can implement the same functionality as either a procedure or a function.

Like tables and views, functions and procedures are saved in the database where you created them. You can set the current database with the `use` command; then, when you define a procedure or function, it will be created in that database.

Now that you've seen how to call functions and procedures, let's look at how to create them.

Creating Functions

Listing 11-1 defines the `f_get_state_population()` function, which accepts a state's name and returns the population of the state.

```
❶ use population;

❷ drop function if exists f_get_state_population;

  delimiter //

❸ create function f_get_state_population(
      state_param    varchar(100)
  )
  returns int
  deterministic reads sql data
  begin
      declare population_var int;

      select  population
      into    population_var
```

```
    from    state_population
    where   state = state_param;

    return(population_var);
❹ end//

delimiter ;
```

Listing 11-1: Creating the `f_get_state_population()` function

In the first line, you set the current database to `population` with the `use` command ❶ so your function will be created in that database.

> **NOTE** *Another way to accomplish this is to specify the database name in the* `create function` *statement by prefixing the function name with the database name and a period, like so:*
>
> ```
> create function population.f_get_state_population(
> ```
>
> *For simplicity's sake, I'll continue using the approach shown in Listing 11-1.*

Before you create the function, you use the `drop function` statement in case there's already a version of this function. If you try to create a function and an old version already exists, MySQL will send a `function already exists` error and won't create the function. Similarly, if you try to drop a function that doesn't already exist, MySQL will also send an error. To prevent that error from appearing, you add `if exists` after `drop function` ❷, which will drop the function if it already exists, but won't send an error if it doesn't.

The function itself is defined between the `create function` ❸ and `end` statements ❹. We'll walk through its components in the following sections.

Redefining the Delimiter

The function definition also includes lines of code to redefine and then reset your delimiter. A *delimiter* is one or more characters that separate one SQL statement from another and mark the end of each statement. Typically, you'll use a semicolon as the delimiter.

In Listing 11-1, you temporarily set the MySQL delimiter to // using the `delimiter //` statement because your function is made up of multiple SQL statements that end in a semicolon. For example, `f_get_state_population()` has three semicolons, located after the `declare` statement, the `select` statement, and the `return` statement. To ensure that MySQL creates your function starting with the `create function` statement and ending with the `end` statement, you need a way to tell it not to interpret any semicolons between those two statements as the end of your function. This is why you've redefined the delimiter.

Let's take a look at what would happen if you didn't redefine your delimiter. If you remove or comment out the `delimiter //` statement at the beginning of your code and look at it in MySQL Workbench, you'll notice some red X markers on lines 12 and 19, indicating errors (Figure 11-1).

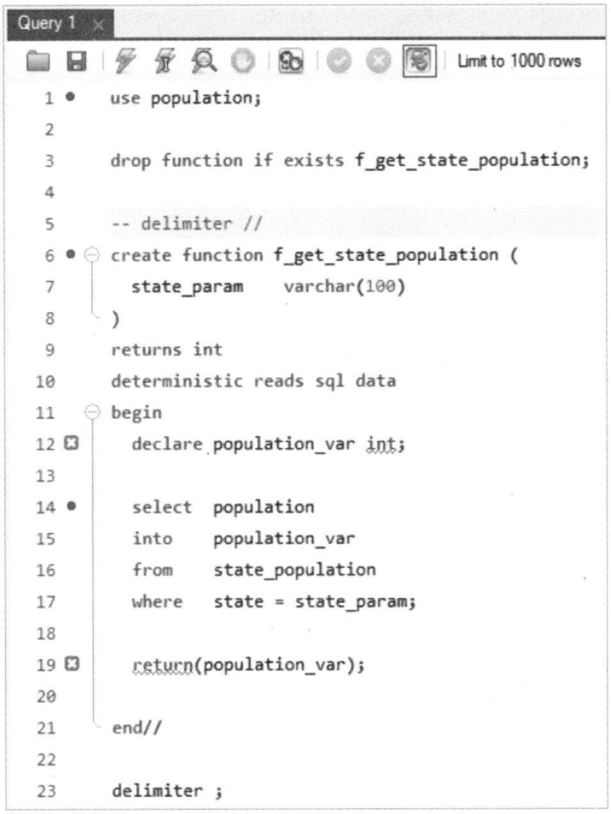

Query 1

Limit to 1000 rows

```
1   use population;
2
3   drop function if exists f_get_state_population;
4
5   -- delimiter //
6   create function f_get_state_population (
7     state_param    varchar(100)
8   )
9   returns int
10  deterministic reads sql data
11  begin
12    declare population_var int;
13
14    select  population
15    into    population_var
16    from    state_population
17    where   state = state_param;
18
19    return(population_var);
20
21  end//
22
23  delimiter ;
```

Figure 11-1: MySQL Workbench showing errors on lines 12 and 19

You commented out the `delimiter` statement on line 5 by adding two hyphens and a space (`-- `) in front of it; this caused MySQL Workbench to report errors on lines 12 and 19 because the semicolon has become the delimiter character. Thus, every time MySQL encounters a semicolon, it assumes that is the end of a SQL statement. MySQL Workbench tries to help you by showing error markers with a red X to let you know the statements ending in semicolons aren't valid.

Redefining the delimiter to `//` (or something other than `;`) informs MySQL Workbench that the statements creating your function aren't over until it hits `//` at the end of line 21. You can fix the errors by uncommenting line 5 (removing the two hyphens and a space, `-- `, at the beginning of the line), thereby reinserting the `delimiter //` command.

After the function has been created, you set the delimiter back to the semicolon on line 23.

NOTE *The typical characters developers use to redefine the delimiter are `//`, `$$`, and occasionally `;;`.*

Although redefining the delimiter to // is necessary here because your function body contains three semicolons, there are other situations where you don't need to redefine your delimiter. For example, you can simplify the following function:

```
delimiter //
create function f_get_world_population()
returns bigint
deterministic no sql
begin
    return(7978759141);
end//

delimiter ;
```

The `begin` and `end` keywords group statements that are part of the function body. Since this function body has only one SQL statement, returning the world population, you don't need to use `begin` and `end` here. And you don't need to redefine your delimiter, either, because there's only one semicolon—at the end of the `return` statement. You can remove the code that redefines and resets the delimiter and simplify your function to this:

```
create function f_get_world_population()
returns bigint
deterministic no sql
return(7978759141);
```

While this is a more concise way to write the function, you might want to keep the `begin` and `end` statements and redefine the delimiter because it makes it easier to add a second SQL statement in the future. The choice is yours.

Adding Parameters and Returning a Value

Both built-in functions and custom functions can accept parameters. You created the `f_get_state_population()` function in Listing 11-1 to accept one parameter named `state_param`, which has a `varchar(100)` data type. You can define parameters with the data types in Chapter 4, including `int`, `date`, `decimal`, and `text`, to define a table's columns.

NOTE *Consider ending any parameters with `_param` to make their role clear.*

Because functions return a value to the caller of the function, you use the `returns` keyword in Listing 11-1 to let MySQL know the data type of the value that your function will return. In this case, the function will return an integer, representing the population of a state.

> **PARAMETERS VS. ARGUMENTS**
>
> Some developers use the words *arguments* and *parameters* interchangeably, although there is a difference between the two. While arguments are the values that are passed to functions, parameters are the variables in the functions that receive those values.
>
> The key point for you, however, is that you can call functions and send values to them, and write functions that receive those values. Functions can be written to accept no values at all, or they can be written to take a large number of them—even thousands. Typically, though, you won't need to write a function that accepts more than 10 parameters.

Specifying Characteristics

In Listing 11-1, once you establish that your function returns an integer, you specify some characteristics of your function. A *characteristic* is an attribute or property of the function. In this example, you used the `deterministic` and `reads sql data` characteristics:

```
deterministic reads sql data
```

You can list the characteristics on one line, or you can list each characteristic on its own line, like so:

```
deterministic
reads sql data
```

You need to choose from two sets of characteristics: `deterministic` or `not deterministic`, and `reads sql data`, `modifies sql data`, `contains sql`, or `no sql`. You must specify at least one of these three characteristics for all of your functions: `deterministic`, `no sql`, or `reads sql data`. If you don't, MySQL will send an error message and won't create your function.

deterministic or not deterministic

Choosing `deterministic` means the function will return the same value given the same arguments and the same state of the database. This is usually the case. The `f_get_state_population()` function is deterministic because, unless the data in the database changes, every time you call `f_get_state_population()` with an argument of `New York`, the function will return the value `19299981`.

The `not deterministic` characteristic means that the function may not return the same value given the same arguments and the same state of the database. This would be the case for a function that returns the current date, for example, as calling it today will yield a different return value than calling it tomorrow.

If you tag a nondeterministic function as `deterministic`, you might get incorrect results when you call your function. If you tag a deterministic function as `not deterministic`, your function might run slower than necessary. If you don't define a function as `deterministic` or `not deterministic`, MySQL defaults to `not deterministic`.

MySQL uses `deterministic` or `not deterministic` for two purposes. First, MySQL has a query optimizer that determines the fastest way to execute queries. Specifying `deterministic` or `not deterministic` helps the query optimizer make good execution choices.

Second, MySQL has a binary log that keeps track of changes to data in the database. The binary log is used to perform *replication*, a process in which data from one MySQL database server is copied to another server, known as a *replica*. Specifying `deterministic` or `not deterministic` helps MySQL perform this replication.

NOTE *Your database administrator can remove the requirement to tag functions as* `deterministic` *or* `not deterministic` *by setting the* `log_bin_trust_function_creators` *configuration variable to* `ON`.

reads sql data, modifies sql data, contains sql, or no sql

The `reads sql data` characteristic means that the function reads from the database using `select` statements but doesn't update, delete, or insert any data; `modifies sql data`, on the other hand, means that the function does update, delete, or insert data. This would be the case more for procedures than for functions because procedures are more commonly used for modifying data in the database than functions are.

The `contains sql` characteristic means the function has at least one SQL statement but doesn't read or write any data from the database, and `no sql` means the function contains no SQL statements. An example of `no sql` would be a function that returns a hardcoded number, in which case it doesn't query the database. You could, for example, write a function that always returns `212` so that you don't need to remember the temperature at which water boils.

If you don't specify `reads sql data`, `modifies sql data`, `contains sql`, or `no sql`, MySQL defaults to `contains sql`.

Defining the Function Body

After listing the characteristics, you define the function body, the block of code that gets executed when the function is called. You use a `begin` and an `end` statement to mark the beginning and end of the function body.

In Listing 11-1, you declared a variable named `population_var` with the `declare` keyword. Variables are named objects that can hold values. You can declare them with any of the MySQL data types; in this case, you used the `int` type. You'll learn about different types of variables in the section "Defining Local Variables and User Variables" later in the chapter.

NOTE *Consider ending your variables with* `_var` *to make their role clear.*

Then you add a select statement that selects the population from your database and writes it into your population_var variable. This select statement is similar to those you've used before, except you're now using the into keyword to select the value you got from the database into a variable.

You then return the value of population_var to the caller of the function with a return statement. Since functions always return one value, there must be a return statement in your function. The data type of the value being returned must match the returns statement at the beginning of the function. You use returns to declare the data type of the value you'll return, and return to actually return the value.

Your end statement is followed by // because you redefined your delimiter to // earlier. Once you reach the end statement, your function body is complete, so you redefine your delimiter back to a semicolon.

> **TRY IT YOURSELF**
>
> **11-1.** In the diet database, the calorie table contains the following data:
>
> ```
> food    calorie_count
> ------  -------------
> banana       110
> pizza        700
> apple        185
> ```
>
> Write a function called f_get_calorie_count() that accepts the name of a food as a parameter and returns the calorie count. The food parameter should be defined as varchar(100). The characteristics should be deterministic and reads sql data.
>
> You can test the function by calling it like so:
>
> ```
> select f_get_calorie_count('pizza');
> ```

Creating Procedures

Similar to functions, procedures accept parameters, include a code block surrounded by begin and end, can have defined variables, and can have a redefined delimiter.

Unlike functions, procedures don't use the returns or return keyword because procedures don't return one value in the way that functions do. Also, you can display values in procedures using the select keyword. Additionally, while MySQL requires you to specify characteristics like deterministic or reads sql data when creating functions, this is not required for procedures.

Listing 11-2 creates a procedure called p_set_state_population() that accepts a parameter for the state's name, gets the latest population values for each county in the state from the county_population table, sums the populations, and writes the total population to the state_population table.

```
❶ use population;

❷ drop procedure if exists p_set_state_population;

❸ delimiter //

❹ create procedure p_set_state_population(
    ❺ in state_param varchar(100)
  )
  begin
    ❻ delete from state_population
      where state = state_param;

    ❼ insert into state_population
      (
            state,
            population
      )
      select state,
          ❽ sum(population)
      from   county_population
      where  state = state_param
      group by state;

❾ end//

  delimiter ;
```

Listing 11-2: Creating the `p_set_state_population()` *procedure*

First, you set your current database to population with use so the procedure will be created in the population database ❶. Before creating the procedure, you check to see if it already exists, and if it does, the old version is deleted with the drop command ❷. Then you redefine your delimiter to // just as you did when creating functions ❸.

Next, you create the procedure and call it p_set_state_population() ❹. As with functions, you name the parameter state_param and give it a varchar(100) data type, and you also specify in to set state_param as an input parameter ❺. Let's look at this step a little closer.

Unlike functions, procedures can accept parameter values as input and also pass values back to the caller as output. They can also accept multiple input and output parameters. (You'll explore output parameters in depth later in this chapter.) When you write procedures, you specify the type of parameter using the keyword in for input, out for output, or inout for parameters that are both. This specification isn't necessary for functions because function parameters are always assumed to be input. If you don't specify in, out, or inout for your procedure parameters, MySQL defaults to in.

Next, the procedure body is between the begin and end statements. In this body, you delete the existing row in the state_population table for the state (if one exists) ❻, then insert a new row into the state_population table ❼. If you don't delete the existing row(s) first, the table will have a row for every time you run the procedure. You want to start with a clean slate before you write the current information to the state_population table.

You get the state's population by summing the populations of the individual counties in that state from the county_population table ❽.

As you did with functions, when you're done defining the procedure you redefine your delimiter to a semicolon ❾.

Using select to Display Values

When you create procedures and functions, you can use the select...into syntax to write a value from the database into a variable. But unlike functions, procedures can also use select statements without the into keyword to display values.

Listing 11-3 creates a procedure called p_set_and_show_state_population() to select the population of the state into a variable and then display a message to the procedure caller.

```
use population;

drop procedure if exists p_set_and_show_state_population;

delimiter //

create procedure p_set_and_show_state_population(
    in state_param varchar(100)
)
begin
  ❶ declare population_var int;

    delete from state_population
    where state = state_param;

  ❷ select sum(population)
    into   population_var
    from   county_population
    where  state = state_param;

  ❸ insert into state_population
    (
           state,
           population
    )
    values
    (
           state_param,
           population_var
    );

  ❹ select concat(
               'Setting the population for ',
               state_param,
```

```
                ' to ',
                population_var
            );
end//

delimiter ;
```

Listing 11-3: Creating the p_set_and_show_state_population() procedure

In this procedure, you declare a variable called population_var as an integer ❶ and insert the sum of the county populations into it using a select...into statement ❷. Then you insert the state_param parameter value and the population_var variable value into your state_population table ❸.

When you call the procedure, it not only sets the correct population of New York in the state_population table, but also displays an informative message:

```
call p_set_and_show_state_population('New York');
```

The message displayed is:

```
Setting the population for New York to 20201249
```

You used select to display the message, which you built by concatenating (using the concat() function), the text Setting the population for, the state_param value, the word to, and the population_var value ❹.

Defining Local Variables and User Variables

The population_var variable is a local variable. *Local variables* are variables that you define in your procedures and functions using the declare command with a data type:

```
declare population_var int;
```

Local variables are only available—or *in scope*—during the execution of the procedure or function containing them. Because you've defined population_var as an int, it will accept only integer values.

You can also use a *user variable*, which starts with the at sign (@) and can be used for the entire length of your session. As long as you're connected to your MySQL server, the user variable will be in scope. If you create a user variable from MySQL Workbench, for example, it will be available until you close the tool.

When creating a local variable, you must specify its data type; when creating a user variable, it's not necessary.

You might see code in a function or procedure that uses both local variables and user variables:

```
declare local_var int;
set local_var = 2;
set @user_var = local_var + 3;
```

You never declared the `@user_var` variable with a data type like `int`, `char`, or `bool`, but because it was being set to an integer value (the `local_var` value plus 3), MySQL automatically set it to `int` for you.

FUN WITH USER VARIABLES

Create a function in the `weird_math` database called `f_math_trick()`:

```
use weird_math;

drop function if exists f_math_trick;

delimiter //

create function f_math_trick(
    input_param   int
)
returns int
no sql
begin
    set @a = input_param;
    set @b = @a * 3;
    set @c = @b + 6;
    set @d = @c / 3;
    set @e = @d - @a;

    return(@e);
end//

delimiter ;
```

The function takes an integer parameter and returns an integer value. You're using several user variables—`@a`, `@b`, `@c`, `@d`, and `@e`—to perform mathematical calculations based on the value of the input argument. The function takes the `input_param` parameter value and utilizes user variables to multiply it by 3, add 6, divide by 3, and subtract the parameter value. At the end of the function, you return the value of the `@e` user variable.

You can run the function like so:

```
select f_math_trick(12);
```

The result is:

```
f_math_trick(12)
----------------
       2
```

(continued)

You passed the f_math_trick() function an argument of 12, and the function returned 2. Try testing other values by calling the function three times with the arguments -28, 0, and 175.

```
select f_math_trick(-28),
       f_math_trick(0),
       f_math_trick(175);

f_math_trick(-28)    f_math_trick(0)    f_math_trick(175)
-----------------    ---------------    -----------------
        2                   2                   2
```

No matter what value you send as an argument to this function, it always returns 2!

Using Logic in Procedures

In procedures, you can use similar programming logic to what you'd use in programming languages like Python, Java, or PHP. For example, you can control the flow of execution with conditional statements like if and case to execute parts of your code under specific conditions. You can also use loops to repeatedly execute parts of your code.

if Statements

An if statement is a decision-making statement that executes particular lines of code if a condition is true. Listing 11-4 creates a procedure called p_compare_population() that compares the population in the state_population table to the county_population table. If the population values match, it returns one message. If they don't, it returns another.

```
use population;

drop procedure if exists p_compare_population;

delimiter //

create procedure p_compare_population(
    in state_param varchar(100)
)
begin
    declare state_population_var int;
    declare county_population_var int;

    select  population
 ❶ into    state_population_var
    from    state_population
    where   state = state_param;
```

```
    select  sum(population)
❷   into    county_population_var
    from    county_population
    where   state = state_param;

❸   if (state_population_var = county_population_var) then
        select 'The population values match';
❹   else
        select 'The population values are different';
    end if;

end//

delimiter ;
```

Listing 11-4: The p_compare_population() procedure

In the first select statement, you select the population for the state from the state_population table and write it into the state_population_var variable ❶. Then, in the second select statement, you select the sum of the populations for each county in the state from the county_population table and write it into the county_population_var variable ❷. You compare the two variables with the if...then syntax. You're saying if the values match ❸, then execute the line that displays the message The population values match; else (otherwise) ❹, execute the next line, displaying the message The population values are different. Then you use end if to mark the end of the if statement.

You call the procedure using the following call statement:

```
call p_compare_population('New York');
```

The result is:

```
The population values are different
```

The procedure shows that the values in the two tables don't match. Perhaps the population table for the counties contains updated data, but the state_population table hasn't been updated yet.

MySQL provides the elseif keyword to check for more conditions. You could expand your if statement to display one of *three* messages:

```
if (state_population_var = county_population_var) then
    select 'The population values match';
elseif (state_population_var > county_population_var) then
    select 'State population is more than the sum of county population';
else
    select 'The sum of county population is more than the state population';
end if;
```

The first condition checks whether the state_population_var value equals the county_population_var value. If that condition is true, the code displays the text The population values match and control flows to the end if statement.

If the first condition was not met, the code checks the elseif condition, which sees if state_population_var is greater than county_population_var. If that condition is true, your code displays the text State population is more than the sum of county population and control flows to the end if statement.

If neither condition is met, control flows to the else statement, the code displays The sum of county population is more than the state population, and control drops down to the end if statement.

case Statements

A case statement is a way to write complex conditional statements. For example, Listing 11-5 defines a procedure that uses a case statement to determine if a state has more than 30 million people, between 10 and 30 million people, or less than 10 million people.

```
use population;

drop procedure if exists p_population_group;

delimiter //

create procedure p_population_group(
    in state_param varchar(100)
)
begin
    declare state_population_var int;

    select population
    into   state_population_var
    from   state_population
    where  state = state_param;

    case
    ❶ when state_population_var > 30000000 then select 'Over 30 Million';
    ❷ when state_population_var > 10000000 then select 'Between 10M and 30M';
    ❸ else select 'Under 10 Million';
    end case;

end//

delimiter ;
```

Listing 11-5: The p_population_group() procedure

Your case statement begins with case and ends with end case. It has two when conditions—which are similar to if statements—and an else statement.

When the condition state_population_var > 30000000 is true, the procedure displays Over 30 Million ❶ and control flows to the end case statement.

When the condition state_population_var > 10000000 is true, the procedure displays Between 10M and 30M ❷ and control flows to the end case statement.

If neither when condition was met, the else statement is executed, the procedure displays Under 10 Million ❸, and control drops down to the end case statement.

You can call your procedure to find out which group a state falls into:

```
call p_population_group('California');
Over 30 Million

call p_population_group('New York');
Between 10M and 30M

call p_population_group('Rhode Island');
Under 10 Million
```

Based on the population retrieved from the database for the state, the case statement displays the correct population grouping for that state.

TRY IT YOURSELF

11-2. The age database contains the following table, called family_member_age, that contains family members' names and ages:

```
person   age
-------  ---
Junior     7
Ricky     16
Grandpa  102
```

Create a procedure called p_get_age_group() that takes a parameter of the family member's name and returns an age group. If the family member is less than 13 years old, the procedure should display the Child age group. If the family member is between 13 and 20 years old, the procedure should display the Teenager age group. Anybody else's age group should appear as Adult.

You can test the procedure like so:

```
call p_get_age_group('Ricky');
```

Loops

You can create *loops* in your procedures to execute parts of your code repeatedly. MySQL allows you to create simple loops, repeat loops, and while loops. This procedure uses a simple loop to display the text Looping Again over and over:

```
drop procedure if exists p_endless_loop;

delimiter //
```

```
create procedure p_endless_loop()
begin
loop
  select 'Looping Again';
end loop;
end;
//
delimiter ;
```

Now call the procedure:

```
call p_endless_loop();
```

You mark the beginning and end of your loop with the `loop` and `end loop` commands. The commands between them will be executed repeatedly.

This procedure displays the text `Looping Again` over and over, theoretically forever. This is called an *endless loop* and should be avoided. You created the loop but didn't provide a way for it to stop. Whoops!

If you run this procedure in SQL Workbench, it opens a different result tab to display the text `Looping Again` each time you go through the loop. Thankfully, MySQL eventually senses that too many result tabs have been opened and gives you the option to stop running your procedure (Figure 11-2).

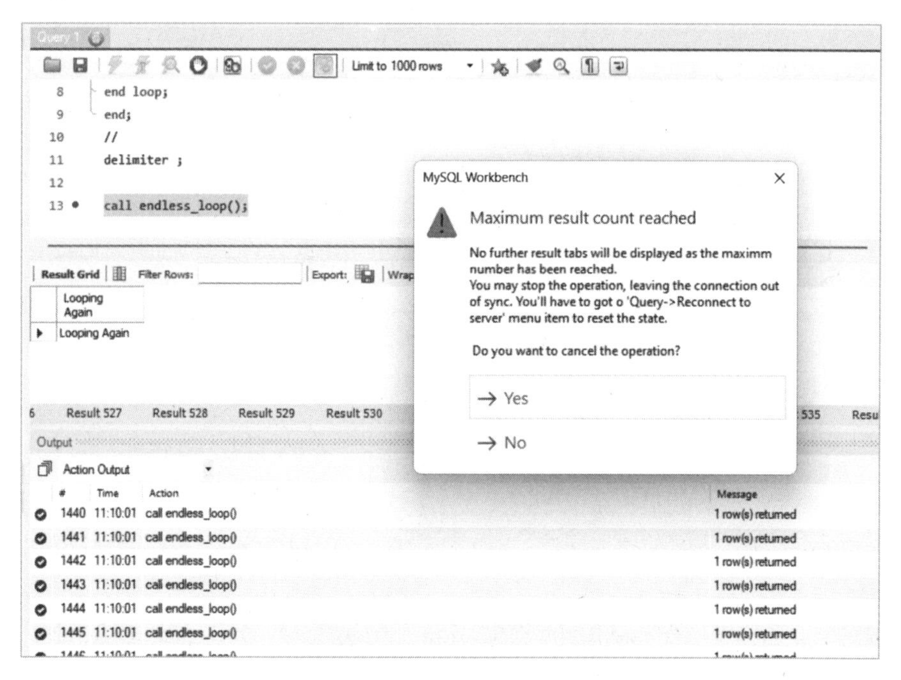

Figure 11-2: Running an endless loop in MySQL Workbench

To avoid creating endless loops, you must design loops to end when some condition has been met. This procedure uses a more sensible simple loop that loops 10 times and then stops:

```
drop procedure if exists p_more_sensible_loop;

delimiter //
create procedure p_more_sensible_loop()
begin
❶ set @cnt = 0;
❷ msl: loop
    select 'Looping Again';
  ❸ set @cnt = @cnt + 1;
  ❹ if @cnt = 10 then
    ❺ leave msl;
    end if;
  end loop msl;
  end;
  //
  delimiter ;
```

In this procedure you define a user variable called `@cnt` (short for *counter*) and set it to `0` ❶. You label the loop `msl` (for *more sensible loop*) by preceding the `loop` statement with `msl:` ❷. Each time you go around in the loop, you add 1 to `@cnt` ❸. In order for the loop to end, the value of `@cnt` must reach `10` ❹. Once it does, you exit the loop using the `leave` command with the name of the loop you want to exit, `msl` ❺.

When you call this procedure, it runs the code between the `loop` and `end loop` statements 10 times, displaying `Looping Again` each time. After the code has been executed 10 times, the loop stops, control drops to the line after the `end loop` statement, and the procedure returns control to the caller.

You can also code a repeat loop with the `repeat...until` syntax, like so:

```
drop procedure if exists p_repeat_until_loop;

delimiter //
create procedure p_repeat_until_loop()
begin
set @cnt = 0;
repeat
   select 'Looping Again';
   set @cnt = @cnt + 1;
until @cnt = 10
end repeat;
end;
//
delimiter ;
```

The code between repeat and end repeat is the body of your loop. The commands between them will be executed repeatedly until @cnt equals 10 and then control will drop down to the end statement. The until statement is at the end of the loop, so the commands in your loop will be executed at least once, because the condition until @cnt = 10 isn't checked until you've gone through the loop the first time.

You can also code a while loop using the while and end while statements:

```
drop procedure if exists p_while_loop;

delimiter //
create procedure p_while_loop()
begin
set @cnt = 0;
while @cnt < 10 do
    select 'Looping Again';
    set @cnt = @cnt + 1;
end while;
end;
//
delimiter ;
```

Your while command specifies the condition that must be met in order for the commands in the loop to be executed. If your condition @cnt < 10 is met, the procedure will do the commands in the loop. When the end while statement is reached, control flows back to the while command and you check again if @cnt is still less than 10. Once your counter is no longer less than 10, control flows to the end command and the loop ends.

Loops are a handy way to repeat functionality when you need to perform similar tasks again and again. Don't forget to give your loops a way to exit so that you avoid writing endless loops, and if you need your loop to execute at least one time, use the repeat...until syntax.

Displaying Procedure Results with select

Since you can use the select statement in procedures, you can write procedures that query data from the database and display the results. When you write a query you'll need to run again, you can save it as a procedure and call the procedure whenever you need it.

Say you wrote a query that selects the populations of all counties in a state, formats them with commas, and orders the counties from largest to smallest. You might want to save your work as a procedure and name it p_get_county_population(), like so:

```
use population;

drop procedure if exists p_get_county_population;
```

```
delimiter //

create procedure p_get_county_population(
    in state_param varchar(100)
)
begin
    select county,
           format(population, 0)
    from   county_population
    where  state = state_param
    order by population desc;
end//

delimiter ;
```

With that procedure in place, you can call it each time you need that information:

```
call p_get_county_population('New York');
```

The results show all 62 counties in New York, with their populations formatted appropriately:

```
Kings          2,736,074
Queens         2,405,464
New York       1,694,251
Suffolk        1,525,920
Bronx          1,472,654
Nassau         1,395,774
Westchester    1,004,457
Erie             954,236
--snip--
```

The next time you want to see the latest version of this data, you can just call the procedure again.

TRY IT YOURSELF

11-3. Create a diet database and a procedure called p_get_food() that takes no parameters. The procedure should display the food and calorie_count columns from the calorie table. Order the results, showing foods with the highest calorie_count value first and the lowest calorie_count last.

You can test the procedure by calling it like so:

```
call p_get_food();
```

Using select in your procedure displays your results. You can also pass values back to the caller of the procedure using the output parameter.

Using a Cursor

While SQL is very good at quickly updating or deleting many rows in a table at once, you'll occasionally need to loop through a dataset and process it one row at a time. You can accomplish this with a cursor.

A *cursor* is a database object that selects rows from the database, holds them in memory, and allows you to loop through them one at a time. To use a cursor, first declare the cursor, then open it, fetch each row from it, and close it. These steps are shown in Figure 11-3.

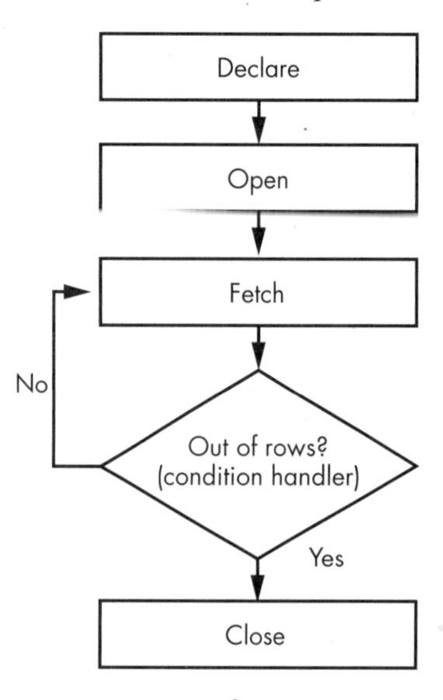

Figure 11-3: Steps for using a cursor

Create a procedure called p_split_big_ny_counties() that uses a cursor. The procedure will use the county_population table, which contains the population of each county within a state. New York has 62 counties, the largest of which are shown here:

```
county       population
-----------  ----------
Kings         2,736,074
Queens        2,405,464
New York      1,694,251
Suffolk       1,525,920
Bronx         1,472,654
Nassau        1,395,774
Westchester   1,004,457
```

Imagine you're a database developer working for the State of New York. You've been asked to break up counties that have over 2 million people into two smaller counties, each containing half of the original county's population.

For example, Kings County has a population of 2,736,074 people. You have been asked to create a county called `Kings-1` with 1,368,037 people and another called `Kings-2` with the remaining 1,368,037. Then you need to delete the original `Kings` row that had a population of 2,736,074. You could write the procedure shown in Listing 11-6 to accomplish this task.

```
drop procedure if exists p_split_big_ny_counties;

delimiter //

create procedure p_split_big_ny_counties()
begin
❶ declare  v_state       varchar(100);
  declare  v_county      varchar(100);
  declare  v_population  int;

❷ declare done bool default false;

❸ declare county_cursor cursor for
    select  state,
            county,
            population
    from    county_population
    where   state = 'New York'
    and     population > 2000000;

❹ declare continue handler for not found set done = true;

❺ open county_cursor;

❻ fetch_loop: loop
    fetch county_cursor into v_state, v_county, v_population;

  ❼ if done then
      leave fetch_loop;
    end if;

  ❽ set @cnt = 1;

  ❾ split_loop: loop

      insert into county_population
      (
        state,
        county,
        population
      )
```

```
      values
      (
        v_state,
        concat(v_county, '-', @cnt),
        round(v_population/2)
      );

      set @cnt = @cnt + 1;

      if @cnt > 2 then
        leave split_loop;
      end if;

    end loop split_loop;

    -- delete the original county
 ❿ delete from county_population where county = v_county;

  end loop fetch_loop;

  close county_cursor;
end;
//

delimiter ;
```

Listing 11-6: Creating the `p_split_big_ny_counties()` *procedure*

This procedure uses a cursor to select the original state, county, and population values from the `county_population` table. You fetch one row at a time from the cursor and loop through your `fetch_loop` once per each row until all the rows have been processed. Let's walk through it.

First you declare the `v_state`, `v_county`, and `v_population` variables that will hold the `state`, `county`, and `population` values for each county with more than 2 million people ❶. You also declare a variable named `done` that will recognize when there are no more rows for your cursor to fetch. You define the `done` variable as a boolean and set its default value to `false` ❷.

Then you declare a cursor, called `county_cursor`, whose `select` statement gets all counties from the `county_population` table that have a population of over 2 million: Kings and Queens counties, in this example ❸.

Next, you declare a condition handler that will automatically set your `done` variable to `true` when your cursor has no more rows to read ❹. *Condition handlers* define how MySQL should respond to situations that arise in your procedures. Your condition handler handles the condition `not found`; if no more rows are found by your `fetch` statement, the procedure will execute the `set done = true` statement, which will change the value of your `done` variable from `false` to `true`, letting you know that there are no more rows to fetch.

When you declare a condition handler, you can choose for it to `continue`—keep running the procedure—or `exit` after the condition has been handled. You choose `continue` in Listing 11-6.

Next, you `open` the `county_cursor` that you declared earlier to prepare it to be used ❺. You create a `fetch_loop` loop that will fetch and iterate through each row of the county_cursor, one row at a time ❻. After this, you fetch the state, county, and population values for a row from the cursor to your v_state, v_county, and v_population variables.

You check your done variable ❼. If all the rows have been fetched from the cursor, you exit the fetch_loop and control flows to the line after the end loop statement. Then you close your cursor and exit the procedure.

If you aren't done fetching rows from the cursor, set a user variable called @cnt to 1 ❽. Then you enter a loop called split_loop that will do the work of splitting the county into two ❾.

NOTE *Notice that your procedure has nested loops: an outer* fetch_loop *that reads the original county data from the table, and an inner* split_loop *that splits the counties into two smaller counties.*

In split_loop, you insert a row into the county_population table that has a county name with a -1 or -2 appended and a population that is half of the original county's population. The -1 or -2 suffix is controlled by your @cnt user variable. You start @cnt as 1 and each time you loop through the split_loop loop, you add 1 to it. Then you concatenate the original county name to a dash and the @cnt variable. You halve the population by dividing your original population that is saved in the v_population variable by 2.

You can call functions from procedures; for example, you use concat() to add the suffix to the county name and you use round() to make sure the new population value doesn't have a fractional part. If there were an odd number of people in your original county, you wouldn't want the population of the new county to be a number like 1368036.5.

When the @cnt variable is more than 2, your work splitting this county is done, so you leave the split_loop and control flows to the line after your end loop split_loop statement. Then you delete the row for the original county from the database ❿.

You reach the end of your fetch_loop, which concludes your work for this county. Control flows back to the beginning of the fetch_loop where you fetch and begin processing for the next county.

Now you can call your procedure

```
call p_split_big_ny_counties();
```

and then look at the largest counties in New York in the database like so:

```
select *
from   county_population
order by population desc;
```

The results are:

```
state     county       population
--------  -----------  ----------
New York  New York       1694251
New York  Suffolk        1525920
New York  Bronx          1472654
```

```
New York  Nassau         1395774
New York  Kings-1        1368037
New York  Kings-2        1368037
New York  Queens-1       1202732
New York  Queens-2       1202732
New York  Westchester    1004457
--snip--
```

Your procedure worked! You now have Kings-1, Kings-2, Queens-1, and Queens-2 counties that are half the size of the original Kings and Queens counties. There are no counties with more than 2 million people, and the original Kings and Queens rows have been removed from the table.

NOTE *Cursors are typically slower than SQL's normal set processing, so when you have a choice of using a cursor or not, it's usually best not to. However, there are times when you need to process each row individually.*

Declaring Output Parameters

So far, all the parameters you've used in your procedures have been input, but procedures also allow you to use output parameters, which pass a value back to the procedure caller. As mentioned earlier, this caller may be a person using a tool like MySQL Workbench, a program written in another programming language like Python or PHP, or another MySQL procedure.

If the caller of the procedure is an end user who just needs to see some values but doesn't need to do any further processing with them, you can use a select statement to display the values. But if the caller needs to use the values, you can pass them back from your procedure as output parameters.

This procedure, called p_return_state_population(), returns the population of a state back to the procedure caller using an output parameter:

```
use population;

drop procedure if exists p_return_state_population;

delimiter //

create procedure p_return_state_population(
  ❶ in  state_param        varchar(100),
  ❷ out current_pop_param   int
)
begin
  ❸ select population
    into   current_pop_param
    from   state_population
    where  state = state_param;
end//

delimiter ;
```

In the procedure, you declare an in (input) parameter named state_param as varchar(100), a string of up to 100 characters ❶. Then you define an out (output) parameter named current_pop_param as an int ❷. You select the population of the state into your output parameter, current_pop_param, which will be automatically returned to the caller because you declared it as an out parameter ❸.

Now call the procedure using a call statement and send New York as an input parameter. Declare that you want the procedure's output parameter to be returned to you as a new user variable called @pop_ny:

```
call p_return_state_population('New York', @pop_ny);
```

The order of the arguments you send to the procedure matches the order of the parameters that you defined when you created the procedure. The procedure was defined to accept two parameters: state_param and current_pop_param. When you call the procedure, you supply the value of New York for the state_param input parameter. Then you supply @pop_ny, which is the name of the variable that will accept the procedure's current_pop_param output parameter value.

You can see the results of the procedure by writing a select statement that displays the value of the @pop_ny variable:

```
select @pop_ny;
```

The result is:

```
20201249
```

The population of New York is saved for you in the @pop_ny user variable.

Writing Procedures That Call Other Procedures

Procedures can call other procedures. For example, here you create a procedure named p_population_caller() that calls p_return_state_population(), gets the value of the @pop_ny variable, and does some additional processing with it:

```
use population;

drop procedure if exists p_population_caller;

delimiter //

create procedure p_population_caller()
begin
  call p_return_state_population('New York', @pop_ny);
  call p_return_state_population('New Jersey', @pop_nj);

  set @pop_ny_and_nj = @pop_ny + @pop_nj;

  select concat(
     'The population of the NY and NJ area is ',
     @pop_ny_and_nj);
```

```
end//

delimiter ;
```

The p_population_caller() procedure calls the p_return_state_population() procedure twice: once with an input parameter of New York, which returns a value to the @pop_ny variable, and once with an input parameter of New Jersey, which returns a value to the @pop_nj variable.

You then create a new user variable called @pop_ny_and_nj and use it to hold the combined populations of New York and New Jersey, by adding @pop_ny and @pop_nj. Then you display the value of the @pop_ny_and_nj variable.

Run your caller procedure using the call statement:

```
call p_population_caller();
```

The result is:

```
The population of the NY and NJ area is 29468379
```

The total population displayed from the caller procedure is 29,468,379, which is the sum of 20,201,249 people in New York and 9,267,130 in New Jersey.

Listing the Stored Routines in a Database

To get a list of the functions and procedures stored in a database, you can query the routines table in the information_schema database:

```
select routine_type,
       routine_name
from   information_schema.routines
where  routine_schema = 'population';
```

The results are:

```
routine_type  routine_name
------------  ------------------------------
FUNCTION      f_get_state_population
PROCEDURE     p_compare_population
PROCEDURE     p_endless_loop
PROCEDURE     p_get_county_population
PROCEDURE     p_more_sensible_loop
PROCEDURE     p_population_caller
PROCEDURE     p_repeat_until_loop
PROCEDURE     p_return_state_population
PROCEDURE     p_set_and_show_state_population
PROCEDURE     p_set_state_population
PROCEDURE     p_split_big_ny_counties
PROCEDURE     p_while_loop
```

As you can see, this query returns a list of functions and procedures in the `population` database.

Summary

In this chapter, you learned to create and call procedures and functions. You used `if` statements, `case` statements, and repeatedly executed functionality with loops. You also saw the benefit of using cursors to process one row at a time.

In the next chapter, you'll create triggers to automatically fire and perform processing for you based on events like rows getting inserted or deleted.

12

CREATING TRIGGERS

In this chapter, you'll create triggers, database objects that automatically *fire*, or execute, before or after a row is inserted, updated, or deleted from a table, and perform the functionality you've defined. Every trigger is associated with one table.

Triggers are most often used to track changes made to a table or to enhance the data's quality before it's saved to the database.

Like functions and procedures, triggers are saved in the database in which you create them.

Triggers That Audit Data

You'll first use triggers to track changes to a database table by creating a second *audit table* that logs which user changed which piece of data and saves the date and time of the change.

Take a look at the following payable table in a company's accounting database.

```
payable_id company            amount   service
----------- -------          -------   ------------------------
     1      Acme HVAC         123.32   Repair of Air Conditioner
     2      Initech Printers 1459.00   Printer Repair
     3      Hooli Cleaning    398.55   Janitorial Services
```

To create an audit table that tracks any changes made to the payable table, enter the following:

```
create table payable_audit
  (
    audit_datetime  datetime,
    audit_user      varchar(100),
    audit_change    varchar(500)
  );
```

You'll create triggers so that when changes are made to the payable table, a record of the changes is saved to the payable_audit table. You'll save the date and time of the change to the audit_datetime column; the user who made the change to the audit_user column; and a text description of what changed to the audit_change column.

Triggers can be set to fire either before or after rows are changed. The first set of triggers you'll create are *after* triggers. You'll set three triggers to fire after changes are made to data in the payable table.

After Insert Triggers

An *after insert* trigger (indicated in the code by the suffix _ai) fires after a row is inserted. Listing 12-1 shows how to create an after insert trigger for the payable table.

```
use accounting;

drop trigger if exists tr_payable_ai;

delimiter //

❶ create trigger tr_payable_ai
  ❷ after insert on payable
  ❸ for each row
  begin
  ❹ insert into payable_audit
    (
      audit_datetime,
      audit_user,
      audit_change
    )
    values
    (
      now(),
```

```
    user(),
    concat(
     'New row for payable_id ',
    ❺ new.payable_id,
      '. Company: ',
      new.company,
      '. Amount: ',
      new.amount,
      '. Service: ',
      new.service
    )
  );
end//

delimiter ;
```

Listing 12-1: Creating an after insert trigger

First you create your trigger and call it tr_payable_ai ❶. Next, you specify the after keyword to indicate when the trigger should fire ❷. In this example, a row will be inserted into the payable table and *then* the trigger will fire, writing the audit row to the payable_audit table.

NOTE *I find it helpful to prefix my triggers with* tr_*, followed by the name of the table I'm tracking, followed by an abbreviation that says when the trigger will fire. These abbreviations include* _bi *(before insert),* _ai *(after insert),* _bu *(before update),* _au *(after update),* _bd *(before delete), and* _ad *(after delete). You'll learn what all these triggers mean throughout this chapter.*

In the trigger, for each row ❸ that gets inserted into the payable table, MySQL will run the code between the begin and end statements. All triggers will include the for each row syntax.

You insert a row into the payable_audit table with an insert statement that calls three functions: now() to get the current date and time; user() to get the username of the user who inserted the row; and concat() to build a string describing the data that was inserted into the payable table ❹.

When writing triggers, you use the new keyword to access the new values being inserted into the table ❺. For example, you got the new payable_id value by referencing new.payable_id, and the new company value by referencing new.company.

Now that you have the trigger in place, try inserting a row into the payable table to see if the new row automatically gets tracked in the payable_audit table:

```
insert into payable
  (
    payable_id,
    company,
    amount,
    service
  )
values
  (
```

```
    4,
    'Sirius Painting',
    451.45,
    'Painting the lobby'
  );

select * from payable_audit;
```

The results show that your trigger worked. Inserting a new row into the payable table caused your tr_payable_ai trigger to fire, which inserted a row into your payable_audit audit table:

```
audit_datetime       audit_user       audit_change
-------------------  --------------   -----------------------------------------
2024-04-26 10:43:14  rick@localhost   New row for payable_id 4.
                                      Company: Sirius Painting. Amount: 451.45.
                                      Service: Painting the lobby
```

The audit_datetime column shows the date and time that the row was inserted. The audit_user column shows the username and the host of the user who inserted the row (the *host* is the server where the MySQL database resides). The audit_change column contains a description of the added row you built with the concat() function.

TRY IT YOURSELF

The jail database has a table called alcatraz_prisoner that contains the following data:

```
prisoner_id  prisoner_name
-----------  -------------
     85      Al Capone
    594      Robert Stroud
   1476      John Anglin
```

12-1. Create an audit table called alcatraz_prisoner_audit in the jail database. Create the table with these columns: audit_datetime, audit_user, and audit_change.

12-2. Write an after insert trigger called tr_alcatraz_prisoner_ai that tracks new rows inserted into the alcatraz_prisoner table to the alcatraz_prisoner_audit table.

You can test the trigger by inserting a new row into alcatraz_prisoner, like so:

```
insert into alcatraz_prisoner
  (
    prisoner_id,
    prisoner_name
  )
```

```
values
  (
    117,
    'Machine Gun Kelly'
  );
```

Check to see that the new row was tracked to the `alcatraz_prisoner_audit` table by selecting from it:

```
select * from alcatraz_prisoner_audit;
```

Do you see an audit row for `Machine Gun Kelly`?

After Delete Triggers

Now you'll write an *after delete* trigger (specified in code with the suffix `_ad`) that will log any rows that are deleted from the `payable` table to the `payable_audit` table (Listing 12-2).

```
use accounting;

drop trigger if exists tr_payable_ad;

delimiter //

create trigger tr_payable_ad
  after delete on payable
  for each row
begin
  insert into payable_audit
    (
      audit_date,
      audit_user,
      audit_change
    )
  values
    (
      now(),
      user(),
      concat(
        'Deleted row for payable_id ',
    ❶ old.payable_id,
        '. Company: ',
        old.company,
        '. Amount: ',
        old.amount,
        '. Service: ',
        old.service
```

```
    )
  );
end//

delimiter ;
```

Listing 12-2: Creating an after delete trigger

The delete trigger looks similar to the insert trigger except for a few differences; namely, you used the old keyword ❶ instead of new. Since this trigger fires when a row is deleted, there are only old values for the columns.

With your after delete trigger in place, delete a row from the payable table and see if the deletion gets logged in the payable_audit table:

```
delete from payable where company = 'Sirius Painting';
```

The results are:

```
audit_datetime       audit_user      audit_change
-------------------  --------------  -------------------------------------------
2024-04-26 10:43:14  rick@localhost  New row for payable_id 4.
                                     Company: Sirius Painting. Amount: 451.45.
                                     Service: Painting the lobby
2024-04-26 10:47:47  rick@localhost  Deleted row for payable_id 4.
                                     Company: Sirius Painting. Amount: 451.45.
                                     Service: Painting the lobby
```

The trigger worked! The payable_audit table still contains the row you inserted into the payable table, but you also have a row that tracked the deletion.

Regardless of whether rows get inserted or deleted, you're logging the changes to the same payable_audit table. You included the text New row or Deleted row as part of your audit_change column value to clarify the action taken.

After Update Triggers

To write an *after update* trigger (_au) that will log any rows that are updated in the payable table to the payable_audit table, enter the code in Listing 12-3.

```
use accounting;

drop trigger if exists tr_payable_au;

delimiter //

create trigger tr_payable_au
  after update on payable
  for each row
begin
❶ set @change_msg =
      concat(
             'Updated row for payable_id ',
```

```
                old.payable_id
        );

❷ if (old.company != new.company) then
    set @change_msg =
          concat(
                @change_msg,
                '. Company changed from ',
                old.company,
                ' to ',
                new.company
          );
  end if;

  if (old.amount != new.amount) then
    set @change_msg =
          concat(
                @change_msg,
                '. Amount changed from ',
                old.amount,
                ' to ',
                new.amount
          );
  end if;

  if (old.service != new.service) then
    set @change_msg =
          concat(
                @change_msg,
                '. Service changed from ',
                old.service,
                ' to ',
                new.service
          );
  end if;

❸ insert into payable_audit
      (
      audit_datetime,
      audit_user,
      audit_change
    )
  values
    (
      now(),
      user(),
      @change_msg
  );

end//

delimiter ;
```

Listing 12-3: Creating an after update trigger

You declare this trigger to fire after an update to the payable table. When you update a row in a table, you can update one or more of its columns. You design your after update trigger to show only the column values that changed in the payable table. For example, if you didn't change the service column, you won't include any text about the service column in the payable_audit table.

You create a user variable called @change_msg ❶ (for *change message*) that you use to build a string that contains a list of every updated column. You check whether each column in the payable table has changed. If the old company column value is different from the new company column value, you add the text Company changed from *old value* to *new value* to the @change_msg variable ❷. You then do the same thing with the amount and service columns, adjusting the message text accordingly. When you're done, the value of @change_msg is inserted into the audit_change column of the payable_audit table ❸.

With your after update trigger in place, see what happens when a user updates a row in the payable table:

```
update payable
set    amount = 100000,
       company = 'House of Larry'
where  payable_id = 3;
```

The first two rows in the payable_audit table are still in the results, along with a new row that tracked the update statement:

```
audit_datetime       audit_user      audit_change
-------------------  --------------  -----------------------------------------
2024-04-26 10:43:14  rick@localhost  New row for payable_id 4.
                                     Company: Sirius Painting. Amount: 451.45.
                                     Service: Painting the lobby
2024-04-26 10:47:47  rick@localhost  Deleted row for payable_id 4.
                                     Company: Sirius Painting. Amount: 451.45.
                                     Service: Painting the lobby
2024-04-26 10:49:20  larry@localhost Updated row for payable_id 3. Company
                                     changed from Hooli Cleaning to House of
                                     Larry. Amount changed from 4398.55 to
                                     100000.00
```

It seems that a user named larry@localhost updated a row, changed the amount to $100,000, and changed the company that will be paid to House of Larry. Hmmm . . .

Triggers That Affect Data

You can also write triggers that fire *before* rows are changed in a table, to change the data that gets written to tables or prevent rows from being inserted or deleted. This can help improve the quality of your data before you save it to the database.

Create a credit table in the bank database that will store customers and their credit scores:

```
create table credit
  (
    customer_id    int,
    customer_name  varchar(100),
    credit_score   int
  );
```

As with after triggers, there are three before triggers that will fire before a row is inserted, deleted, or updated.

Before Insert Triggers

The *before insert* trigger (_bi) fires before a new row is inserted. Listing 12-4 shows how to write a before insert trigger to make sure no scores outside of the 300–850 range (the lowest possible credit score and the highest) get inserted into the credit table.

```
use bank;

delimiter //

❶ create trigger tr_credit_bi
  ❷ before insert on credit
    for each row
  begin
  ❸ if (new.credit_score < 300) then
      set new.credit_score = 300;
    end if;

  ❹ if (new.credit_score > 850) then
      set new.credit_score = 850;
    end if;

  end//

  delimiter ;
```

Listing 12-4: Creating a before insert trigger

First, you name the trigger tr_credit_bi ❶ and define it as a before insert trigger ❷ so that it will fire before rows are inserted into the credit table. Because this is an insert trigger, you can take advantage of the new keyword by checking if new.credit_score—the value about to be inserted into the credit table—is less than 300. If so, you set it to exactly 300 ❸. You do a similar check for credit scores over 850, changing their value to exactly 850 ❹.

Insert some data into the credit table and see what effect your trigger has:

```
insert into credit
  (
    customer_id,
```

```
    customer_name,
    credit_score
  )
values
  (1, 'Milton Megabucks',   987),
  (2, 'Patty Po',           145),
  (3, 'Vinny Middle-Class', 702);
```

Now take a look at the data in the credit table:

```
select * from credit;
```

The result is:

```
customer_id  customer_name       credit_score
-----------  ------------------  ------------
     1       Milton Megabucks        850
     2       Patty Po                300
     3       Vinny Middle-Class      702
```

Your trigger worked. It changed the credit score for Milton Megabucks from 987 to 850 and the credit score for Patti Po from 145 to 300 just before those values were inserted into the credit table.

Before Update Triggers

The *before update* trigger (_bu) fires before a table is updated. You already wrote a trigger that prevents an insert statement from setting a credit score outside of the 300–850 range, but it's possible that an update statement could update a credit score value outside of that range too. Listing 12-5 shows how to create a before update trigger to solve this.

```
use bank;

delimiter //

create trigger tr_credit_bu
  before update on credit
  for each row
begin
  if (new.credit_score < 300) then
    set new.credit_score = 300;
  end if;

  if (new.credit_score > 850) then
    set new.credit_score = 850;
  end if;

end//

delimiter ;
```

Listing 12-5: Creating a before update trigger

Update a row to test your trigger:

```
update credit
set    credit_score = 1111
where  customer_id = 3;
```

Now take a look at the data in the credit table:

```
select * from credit;
```

The result is:

```
customer_id  customer_name       credit_score
-----------  ------------------  ------------
     1       Milton Megabucks        850
     2       Patty Po                300
     3       Vinny Middle-Class      850
```

It worked. The trigger would not let you update the credit score for Vinny Middle-Class to 1111. Instead, it set the value to 850 before updating the row in the table.

TRY IT YOURSELF

The exam database has a table called grade that contains the following data:

```
student_name  score
------------  -----
   Billy        79
   Jane         87
   Paul         93
```

The teacher's policy is that students who scored less than 50 points will be given a score of 50. Students who got all questions correct—including extra credit—cannot get a score higher than 100.

12-3. Write an update trigger called tr_grade_bu that changes scores under 50 to 50, and scores over 100 to 100.

You can test the trigger by updating the scores in the table like so:

```
update grade set score = 38  where student_name = 'Billy';
update grade set score = 107 where student_name = 'Jane';
update grade set score = 95  where student_name = 'Paul';
```

Now check the values in the grade table:

```
select * from grade;
```

(continued)

You should see these results:

```
student_name  score
------------  -----
   Billy        50
   Jane        100
   Paul         95
```

The trigger should have set Billy's score to `50`, Jane's to `100`, and Paul's to `95`.

Before Delete Triggers

Lastly, a *before delete* trigger (`_bd`) will fire before a row is deleted from a table. You can use a before delete trigger as a check before you allow the row to be deleted.

Say your bank manager asked you to write a trigger that prevents users from deleting any customers from the `credit` table that have a credit score over 750. You can achieve this by writing a before delete trigger as shown in Listing 12-6.

```
use bank;

delimiter //

create trigger tr_credit_bd
  before delete on credit
  for each row
begin
❶ if (old.credit_score > 750) then
    signal sqlstate '45000'
    set message_text = 'Cannot delete scores over 750';
  end if;
end//

delimiter ;
```

Listing 12-6: Creating a before delete trigger

If the credit score of the row you're about to delete is over 750, the trigger returns an error ❶. You use a `signal` statement, which handles returning an error, followed by the `sqlstate` keyword and code. A *sqlstate code* is a five-character code that identifies a particular error or a warning. Since you're creating your own error, you use `45000`, which represents a user-defined error. Then, you define the `message_text` to display your error message.

Test your trigger by deleting some rows from the `credit` table:

```
delete from credit where customer_id = 1;
```

Since customer `1` has a credit score of 850, the result is:

```
Error Code: 1644. Cannot delete scores over 750
```

Your trigger worked. It prevented the deletion of the row because the credit score was over 750.

Now delete the row for customer `2`, who has a credit score of 300:

```
delete from credit where customer_id = 2;
```

You get a message back informing you that the row was deleted:

```
1 row(s) affected.
```

Your trigger is working as you intended. It allowed you to delete the row for customer `2` because their credit score was not more than 750, but prevented you from deleting customer `1` because their credit score was over 750.

Summary

In this chapter, you created triggers that automatically fire and perform tasks you define. You learned the differences between before and after triggers, and the three types of each. You used triggers to track changes to tables, prevent particular rows from being deleted, and control ranges of allowed values.

In the next chapter, you'll learn how to use MySQL events to schedule tasks.

13

CREATING EVENTS

In this chapter, you'll create *events*. Also called scheduled events, these are database objects that fire based on a set schedule, executing the functionality you defined when creating them.

Events can be scheduled to run once or at some interval, like daily, weekly, or yearly; for example, you might create an event to perform weekly payroll processing. You can use events to schedule long-running processing during off-hours, like updating a billing table based on orders that came in that day. Sometimes you schedule off-hour events because your functionality needs to happen at a particular time, like making changes to the database at 2 AM when Daylight Saving Time begins.

The Event Scheduler

MySQL has an *event scheduler* that manages the scheduling and execution of events. The event scheduler can be turned on or off, but should be on by default. To confirm that the scheduler is on, run the following command:

```
show variables like 'event_scheduler';
```

If your scheduler is on, the result should look as follows:

```
Variable_name    Value
---------------  -----
event_scheduler   ON
```

If the `Value` displayed is `OFF`, you (or your database administrator) need to turn the scheduler on with this command:

```
set global event_scheduler = on;
```

If the `Value` returned is `DISABLED`, your MySQL server was started with the scheduler disabled. Sometimes this is done to temporarily stop the scheduler. You can still schedule events, but no events will fire until the scheduler is enabled again. If the event scheduler is disabled, it needs to be changed in a configuration file managed by your database administrator.

Creating Events with No End Date

In Listing 13-1 you create an event that removes old rows from the `payable_audit` table in the `bank` database.

```
use bank;

drop event if exists e_cleanup_payable_audit;

delimiter //

❶ create event e_cleanup_payable_audit
  ❷ on schedule every 1 month
  ❸ starts '2024-01-01 10:00'
❹ do
  begin
  ❺ delete from payable_audit
    where audit_datetime < date_sub(now(), interval 1 year);
  end //

delimiter ;
```

Listing 13-1: Creating a monthly event

To create the event in the bank database, first you set your current database to bank with the use command. Then you drop the old version of this event (if one exists) in order to create a new one. Next you create the event, e_cleanup_payable_audit ❶, and set a schedule to run it once per month.

NOTE *Consider beginning events with e_ to make their purpose clear.*

Every event begins with on schedule; for a one-time event, you'd follow this with the at keyword and the timestamp (the date and time) at which the event should fire. For a recurring event, on schedule should be followed by the word every and the interval at which it should fire. For example, every 1 hour, every 2 week, or every 3 year. (Intervals are expressed in the singular form, like 3 year and not 3 years.) In this case, you specify every 1 month ❷. You'll also define the date and time when the recurring event starts and ends.

For your event, you define starts as 2024-01-01 10:00 ❸, meaning your event will start firing on 1/1/2024 at 10 AM and will fire every month at this time. You didn't use the ends keyword, so this event will fire monthly—theoretically forever—until the event is dropped with the drop event command.

Then, you define the event's actions with the do command ❹, and add the SQL statements that perform the functionality in the event body. Your event body starts with begin and ends with end. Here, you delete rows in the payable_audit table that are more than one year old ❺. While you use only one statement here, it is possible to put multiple SQL statements in the event body.

The show events command displays a list of scheduled events in the current database, as in Figure 13-1.

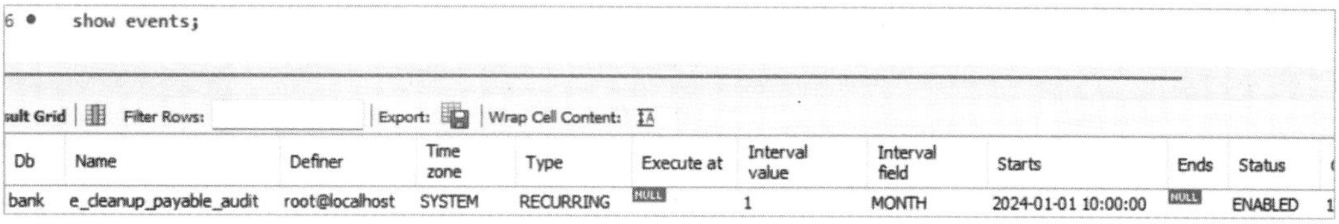

```
6 •    show events;
```

Db	Name	Definer	Time zone	Type	Execute at	Interval value	Interval field	Starts	Ends	Status
bank	e_cleanup_payable_audit	root@localhost	SYSTEM	RECURRING	NULL	1	MONTH	2024-01-01 10:00:00	NULL	ENABLED

Figure 13-1: The show events command as seen in MySQL Workbench

The user account that defined the event is listed as the Definer. This gives you an audit trail that tells you who scheduled which events.

To show only events for a particular database (even if you aren't currently in that database), use the show events in *database* command. In this example, the command would be show events in bank.

To get a list of all events in all databases, you can use the following query:

```
select * from information_schema.events;
```

MySQL provides you with the events table in the information_schema database that you can query for this purpose.

Creating Events with an End Date

For events that should run for a limited time, use the ends keyword. For example, you might want to create an event that runs at 1/1/2024 once an hour between 9 AM and 5 PM:

```
on schedule every 1 hour
starts '2024-01-01 9:00'
ends '2024-01-01 17:00'
```

To schedule an event that runs every 5 minutes for the next hour, you might enter the following:

```
on schedule every 5 minute
starts current_timestamp
ends current_timestamp + interval 1 hour
```

You started your event immediately. It will fire every 5 minutes, and will stop firing one hour from now.

Sometimes you need an event to fire just once at a particular date and time. For example, you may need to wait until after midnight to do some one-time account updates to your bank database so that interest rates are calculated first by another process. You could define an event like so:

```
use bank;

drop event if exists e_account_update;

delimiter //

create event e_account_update
on schedule at '2024-03-10 00:01'
do
begin
  call p_account_update();
end //

delimiter ;
```

Your e_account_update event is scheduled to run on 3/10/2024 at 1 minute past midnight.

NOTE *Events can call procedures. In this example, you moved the functionality that updates your account out of the event and into the* p_account_update() *procedure, which allows you to call it from your scheduled event but also call the procedure directly to execute it immediately.*

You might find it useful to schedule a one-time event when the clocks change to Daylight Saving Time. On 3/10/2024, for example, the clocks move forward one hour. On 11/6/2024, Daylight Saving Time ends and the clocks move back one hour. In many databases, data will need to change as a result.

Schedule a one-time event for March 10, 2024, so that the database makes changes when Daylight Saving Time begins. On that date at 2 AM, your system clock will change to 3 AM. Schedule your event for 1 minute before the clocks change:

```
use bank;

drop event if exists e_change_to_dst;

delimiter //

create event e_change_to_dst
on schedule
at '2024-03-10 1:59'
do
begin
  -- Make any changes to your application needed for DST
  update current_time_zone
  set    time_zone = 'EDT';
end //

delimiter ;
```

Rather than having to stay awake until 1:59 in the morning to change the clock, you can schedule an event to do it for you.

Checking for Errors

To check for errors after your event runs, query a table in the `performance_schema` database called `error_log`.

The `performance_schema` database is used to monitor the performance of MySQL. The `error_log` table houses diagnostic messages like errors, warnings, and notifications of the MySQL server starting or stopping.

For example, you can check all event errors by finding rows where the `data` column contains the text `Event Scheduler`:

```
select *
from   performance_schema.error_log
where  data like '%Event Scheduler%';
```

This query finds all rows in the table that have the text `Event Scheduler` somewhere in the `data` column. Recall from Chapter 7 that the `like` operator allows you to check if a string matches some pattern. Here you're using the `%` wildcard character to check that the `data` column contains a value that starts with any character(s), contains the text `Event Scheduler`, then ends with any character(s).

To find errors for a particular event, search for the event name. Say the e_account_update event calls a procedure named p_account_update(), but that procedure doesn't exist. You'll find errors for the e_account_update event like so:

```
select  *
from    performance_schema.error_log
where   data like '%e_account_update%';
```

The query returns a row that shows the logged column with the date and time when the event fired, and the data column shows an error message (Figure 13-2).

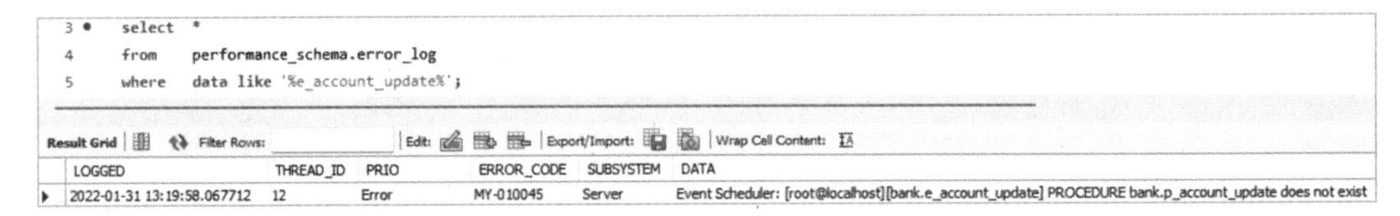

Figure 13-2: Displaying event errors in MySQL Workbench

The message tells you that the e_account_update event in the bank database failed because p_account_update does not exist.

You can disable an event using the alter command:

```
alter event e_cleanup_payable_audit disable;
```

The event will not fire again until you re-enable it, like so:

```
alter event e_cleanup_payable_audit enable;
```

When an event is no longer needed, you can drop it from the database using the drop event command.

TRY IT YOURSELF

13-1. Create a recurring event in the eventful database called e_write_timestamp that starts firing now and stops firing in 5 minutes. Create the event so that every minute, the event writes the current timestamp into the message column of the event_message table, using this command:

```
insert into event_message (message)
values (current_timestamp);
```

13-2. Check if there were any errors for the event.

13-3. Over the next 5 minutes, check the contents of the event_message table using the select * from event_message; command. Are new timestamps being inserted into the table every minute?

Summary

In this chapter, you scheduled events to fire once and on a recurring basis. You learned how to check for errors in your event scheduler, and disable and drop events. The next chapter will focus on assorted tips and tricks that can make MySQL more productive and enjoyable.

PART IV

ADVANCED TOPICS

In Part IV, you'll learn how to load data to and from files, run MySQL commands from script files, avoid common pitfalls, and use MySQL within programming languages.

In Chapter 14, we'll go over some tips and tricks for avoiding common problems, and you'll see how to load data to or from a file. You'll also take a look at using transactions and the MySQL command line client.

In Chapter 15, you'll use MySQL from programming languages like PHP, Python, and Java.

14

TIPS AND TRICKS

In this chapter, you'll build confidence in your new MySQL skills by reviewing common pitfalls and how to avoid them. Then, you'll look at transactions and the MySQL command line client. You'll also learn how to load data to and from files.

Common Mistakes

MySQL can process sets of information very quickly. You can update thousands of rows in the blink of an eye. While this gives you a lot of power, it also means there is greater potential for mistakes, like running SQL against the wrong database or server or running partial SQL statements.

Working in the Wrong Database

When working with relational databases like MySQL, you need to be cognizant of which database you're working in. It's surprisingly common to run a SQL statement in the wrong one. Let's look at some ways you can avoid it.

Say you've been asked to create a new database called `distribution` and to create a table called `employee`.

You might use these SQL commands:

```
create database distribution;

create table employee
  (
    employee_id     int             primary key,
    employee_name   varchar(100),
    tee_shirt_size  varchar(3)
  );
```

If you're using MySQL Workbench to run the commands, you'll see two green checkmarks in the lower panel telling you that your database and table were successfully created (Figure 14-1).

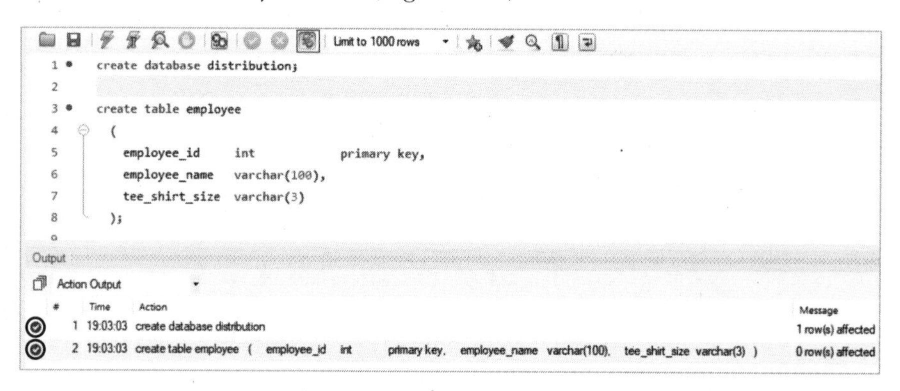

Figure 14-1: You used MySQL Workbench to create an `employee` table in the `distribution` database . . . didn't you?

Everything looks good, so you declare victory and move on to your next task. Then you start to get calls saying that the table wasn't created. What went wrong?

Although you created the `distribution` database, you didn't set your current database to `distribution` before you created the table. Your new `employee` table instead was created in whatever your current database happened to be at the time. You should have included the `use` command before you created the table, like so:

```
create database distribution;

use distribution;

create table employee
```

```
(
  employee_id     int            primary key,
  employee_name   varchar(100),
  tee_shirt_size  varchar(3)
);
```

One way to avoid creating a table in the wrong database is to *fully qualify* the table name when you create it. You can specify the name of the database to create the table in, so that even if you aren't currently in that database, the table will be created there. Here you specify that you want to create the employee table in the distribution database:

```
create table distribution.employee
```

Another way you could have avoided creating the table in the wrong database is by checking what your current database was before creating the table, like so:

```
select database();
```

If your result was anything other than distribution, this would have alerted you that you forgot to correctly set your current database with the use command.

You can fix such a mistake by figuring out which database the employee table was created in, dropping the table, and then re-creating the employee table in the distribution database.

To determine which database or databases have an employee table, run this query:

```
select table_schema,
       create_time
from   information_schema.tables
where  table_name = 'employee';
```

You've queried the tables table in the information_schema database, and selected the create_time column to see if this table was created recently. The output is as follows:

```
TABLE_SCHEMA  CREATE_TIME
------------  -------------------
     bank     2024-02-05 14:35:00
```

It's possible that you could have different tables named employee in more than one database. If that were the case, your query would have returned more than one row. But in this example, the only database with an employee table is bank, so that's where your table was mistakenly created.

As an extra check, see how many rows are in the employee table:

```
use bank;

select count(*) from employee;
```

```
count(*)
--------
    0
```

There are no rows in this table, which is expected for a table that was created by mistake. Confident that the `employee` table in the `bank` database is the one you accidentally created in the wrong place, you can now run these commands to correct your mistake:

```
use bank;

-- Remove the employee table mistakenly created in the bank database
drop table employee;

use distribution;

-- Create the employee table in the bank database
create table employee
  (
    employee_id     int            primary key,
    employee_name   varchar(100),
    tee_shirt_size  varchar(3)
  );
```

Your `employee` table in the `bank` database has been removed, and an `employee` table in the `distribution` database has been created.

You could have moved the table from one database to the other with the `alter table` command instead, like so:

```
alter table bank.employee rename distribution.employee;
```

It's preferable to drop and re-create a table rather than altering the table, however, especially if the table has triggers or foreign keys associated with it that might still be pointing to the wrong database.

Using the Wrong Server

Sometimes, SQL statements can be executed against the wrong MySQL server. Companies often set up different servers for production and development. The *production* environment is the live environment that end users access, so you want to be careful with its data. The *development* environment is where developers test new code. Since the data in this environment is seen only by developers, you should always test your SQL statements here before releasing them to production.

It's not unusual for a developer to have two windows open: one connected to the production server and one connected to the development server. If you're not careful, you can make changes in the wrong window.

If you're using a tool like MySQL Workbench, consider naming your connections *Production* and *Development* so that its tabs clearly state which environment is which (see Figure 14-2).

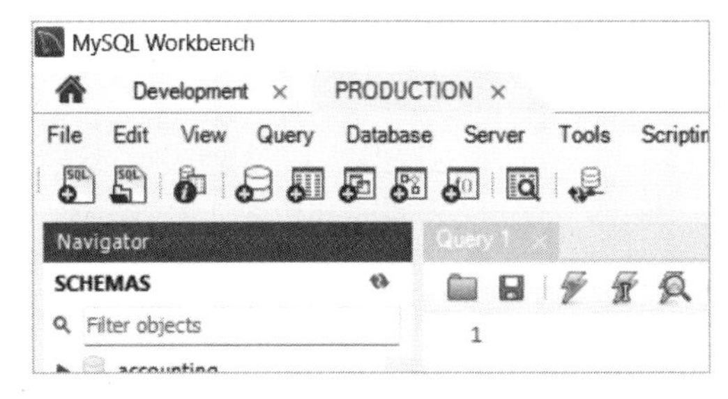

Figure 14-2: Naming the MySQL Workbench tabs Development and Production

To name connections in MySQL Workbench, go to **Database ▸ Manage Connections**. In the Setup New Connection window that opens, enter a connection name like Development or Production that specifies the environment.

NOTE *To avoid potential mistakes, consider opening a connection to the production environment* only *when you're making changes that have already been tested, and closing that window once you're finished.*

Other tools have similar ways to mark production and development environments. Some allow you to change the background color, so you might consider setting your production screen to red as a reminder to be careful in that environment.

Leaving where Clauses Incomplete

When you insert, update, or delete data in a table, it's crucial that your `where` clause is complete. If it isn't, you run the risk of changing unintended rows.

Imagine you own a used car dealership and you store the cars you have in stock in the `inventory` table. Check what's in the table:

```
select * from inventory;
```

The results are:

```
vin                mfg        model    color
----------         ---------- -------- ------
1ADCQ67RFGG234561  Ford       Mustang  red
2XBCE65WFGJ338565  Toyota     RAV4     orange
3WBXT62EFGS439561  Volkswagen Golf     black
4XBCX68RFWE532566  Ford       Focus    green
5AXDY62EFWH639564  Ford       Explorer yellow
6DBCZ69UFGQ731562  Ford       Escort   white
7XBCX21RFWE532571  Ford       Focus    black
8AXCL60RWGP839567  Toyota     Prius    gray
9XBCX11RFWE532523  Ford       Focus    red
```

Looking at a Ford Focus on the lot, you notice that you have it listed as green, but its color is actually closer to blue. You decide to update its color in the database (Listing 14-1).

```
update inventory
set    color = 'blue'
where  mfg = 'Ford'
and    model = 'Focus';
```

Listing 14-1: An update statement with incomplete criteria in the `where` *clause*

When you run the `update` statement, you're surprised to see MySQL return a message of `3 row(s) affected`. You meant to update only one row, but it appears that three rows were changed.

You run a query to see what happened:

```
select *
from   inventory
where  mfg = 'Ford'
and    model = 'Focus';
```

The results are:

```
4XBCX68RFWE532566  Ford  Focus  blue
7XBCX21RFWE532571  Ford  Focus  blue
9XBCX11RFWE532523  Ford  Focus  blue
```

Because the `where` clause in your `update` statement was missing criteria, you mistakenly updated the color of every Ford Focus in the table to `blue`.

Your `update` statement in Listing 14-1 should have been:

```
update inventory
set    color = 'blue'
where  mfg = 'Ford'
and    model = 'Focus'
and    color = 'green';
```

The last line is missing in Listing 14-1. With this additional criteria, the `update` statement would have changed just *green* Ford Focuses to blue. Since you had only one green Ford Focus on the lot, only the correct car would have been updated.

A more efficient way to do the update would have been to use the VIN (Vehicle Identification Number) in your `where` clause:

```
update inventory
set    color = 'blue'
where  vin = '4XBCX68RFWE532566';
```

Since each car has a distinct VIN, with this approach you are guaranteed that your `update` statement will update just one vehicle.

Either of these update statements would have provided enough criteria to identify the one row you intended to change, and you would have updated just that row.

A simple sanity check you can perform before you insert, update, or delete rows is to select from the table using the same where clause. For example, if you were planning to run the update statement in Listing 14-1, you would run this select statement first:

```
select *
from   inventory
where  mfg = 'Ford'
and    model = 'Focus';
```

The results would have been:

```
vin                  mfg         model     color
----------           ----------  --------  ------
4XBCX68RFWE532566    Ford        Focus     green
7XBCX21RFWE532571    Ford        Focus     black
9XBCX11RFWE532523    Ford        Focus     red
```

The query produces a list of the rows that you are about to update. If you really wanted to update all three rows, you could then run the update statement that uses the same where clause. In this case, you would have recognized that this where clause in your select statement matched too many rows and could have avoided updating more than the single row you intended.

Running Partial SQL Statements

MySQL Workbench has three lightning bolt icons that can be used for executing SQL statements in different ways. Each icon's actions are listed in Table 14-1.

Table 14-1: Lightning Bolt Icons in MySQL Workbench

Simple lightning bolt	Executes the selected statements or, if nothing is selected, all statements
Cursor lightning bolt	Executes the statement under the keyboard cursor
Magnifying glass lightning bolt	Executes the EXPLAIN plan for the statement under the cursor

Most MySQL Workbench users will use the simple and cursor lightning bolt icons for their day-to-day work. The magnifying glass lightning bolt is used less often, as it is an optimization tool that explains what steps MySQL would take to run your query.

If you use the simple lightning bolt icon without realizing part of your SQL statement is highlighted, you'll inadvertently run that highlighted section. For example, say you want to delete the Toyota Prius from the `inventory` table. You write the following `delete` statement to delete the car with the Prius's VIN:

```
delete from inventory
where vin = '8AXCL60RWGP839567';
```

Now, you'll use MySQL Workbench to run your `delete` statement (Figure 14-3).

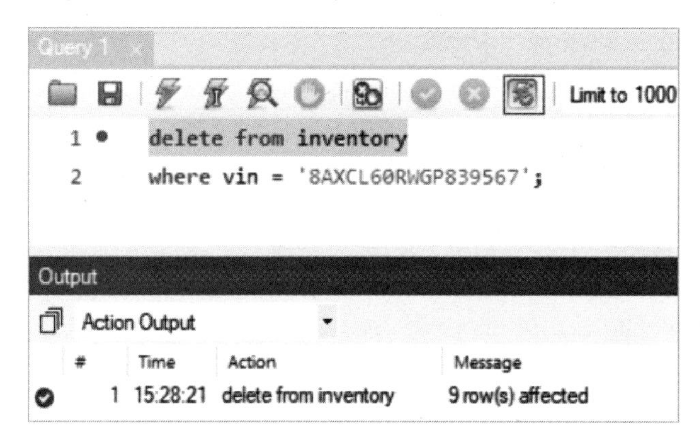

Figure 14-3: Mistakenly deleting all rows in the table using MySQL Workbench

When you click the simple lightning bolt icon, MySQL tells you that all rows in the table were deleted. What happened?

Before you ran the `delete` statement, you mistakenly highlighted the first line of the SQL command. This told MySQL to delete all rows in the table, rather than the one you attempted to specify.

Transactions

You can lessen the possibility of mistakes by executing statements as part of a transaction. A *transaction* is a group of one or more SQL statements that can be *committed* (made permanent) or *rolled back* (canceled). For example, before updating your `inventory` table, you could use the `start transaction` command to begin a transaction that you can later commit or roll back:

```
start transaction;

update inventory
set    color = 'blue'
where  mfg = 'Ford'
and    model = 'Focus';
```

The `begin` command is an alias for `start transaction`. You can use either.

If you run your update statement and MySQL returns a message of 3 row(s) affected, but you were expecting one row to be changed, you can roll back the transaction:

```
rollback;
```

Your update gets rolled back, the changes are canceled, and the rows remain unchanged in the table. To make the changes permanent, commit the transaction:

```
commit;
```

Transactions are helpful when you're using Data Manipulation Language (DML) statements like insert, update, or delete. Data Definition Language (DDL) statements like create function, drop procedure, or alter table shouldn't be made in a transaction. They can't be rolled back—running them will automatically commit the transaction.

TRY IT YOURSELF

The zoo table in the travel database contains the following data:

```
zoo_name                 country
-----------------------  ---------
Beijing Zoo              China
Berlin Zoological Garden Germany
Bronx Zoo                USA
Ueno Zoo                 Japan
Singapore Zoo            Singapore
Chester Zoo              England
San Diego Zoo            USA
Toronto Zoo              Canada
Korkeasaari Zoo          Finland
Henry Doorly Zoo         USA
```

14-1. Using MySQL Workbench, run these commands one at a time:

```
use travel;

start transaction;

update zoo
set    zoo_name = 'SD Zoo'
where  country = 'USA';

rollback;

select * from zoo;
```

(continued)

Did the update statement take effect, or was it rolled back?

14-2. Now run the same commands one at a time, but instead of using rollback, use commit. Does the update statement take effect now?

```
use travel;

start transaction;

update zoo
set    zoo_name = 'SD Zoo'
where  country = 'USA';

commit;

select * from zoo;
```

Did the update statement update only the San Diego Zoo? Should this update statement have been rolled back or committed?

Until you commit or roll back your update statement, MySQL will keep the table locked. For example, if you run these commands

```
start transaction;

update inventory
set    color = 'blue'
where  mfg = 'Ford'
and    model = 'Focus';
```

the inventory table will remain locked and no other users will be able to make changes to its data until you commit or roll back your changes. If you start the transaction and then go to lunch without committing or rolling it back, you might come back to some angry database users.

Supporting an Existing System

You may find yourself supporting a MySQL system that has already been developed. A good way to start to understand an existing system is by browsing through its database objects using MySQL Workbench (Figure 14-4).

You can learn a lot about an existing system by exploring using MySQL's navigator panel. Are there many databases with a few tables in each one, or are there one or two databases with a lot of tables in each? What are the naming conventions you should follow? Are there many stored procedures, or is most of the business logic handled outside of MySQL in a programming language like PHP or Python? Have primary and foreign keys been set up for most tables? Do they use many triggers? Looking at the procedures, functions, and triggers, which delimiter do they use? Check the existing database objects and follow that lead when it comes to naming conventions for any new code you add to the system.

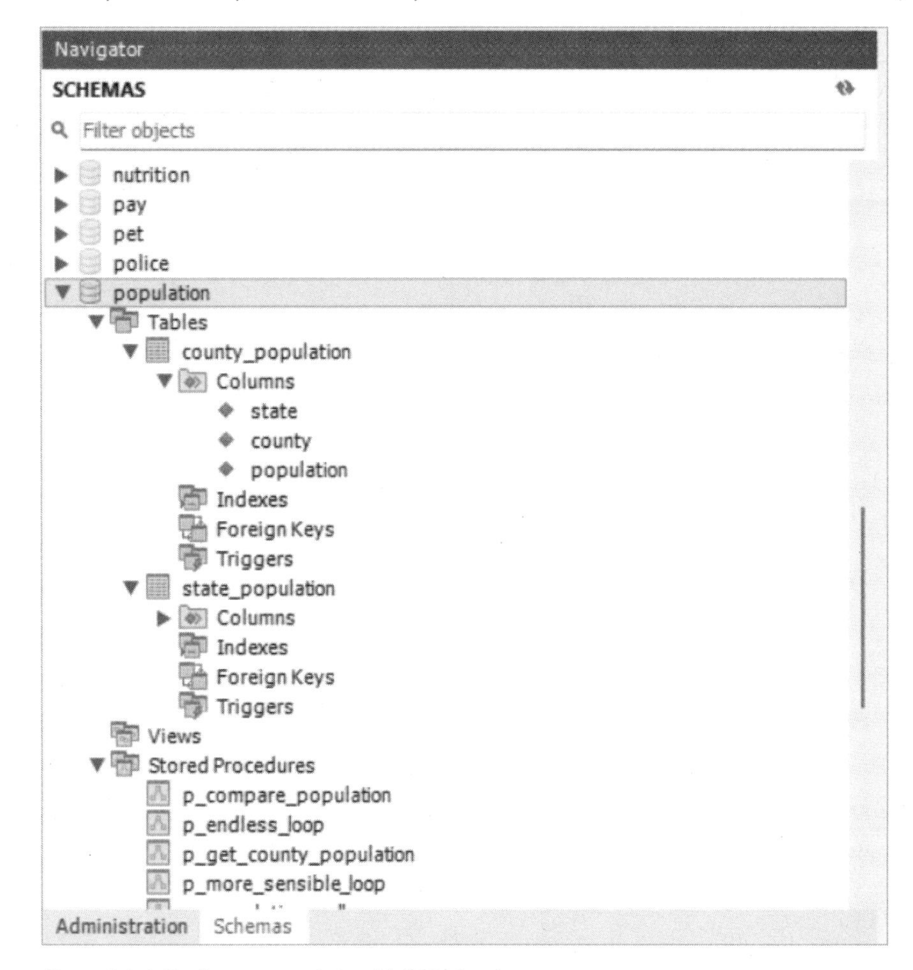

Figure 14-4: Exploring an existing MySQL database

Sometimes the hardest part of supporting an existing system is understanding the problem set and terminology of the application. A good first question for you to ask is, "What are the most important tables?" Focus your attention on learning about them first. You can select from the table and understand how those primary key values uniquely identify the rows in the table. Check for triggers on those tables and look through the trigger code to understand what actions happen automatically when data in the tables is changed.

MySQL Workbench also presents a nice graphical depiction of the way that MySQL objects hang together. For example, you can see in Figure 14-4 that databases have tables, stored procedures, and functions. Tables have columns, indexes, foreign keys, and triggers.

Using the MySQL Command Line Client

The MySQL command line client, mysql, allows you to run SQL commands from the command line interface of your computer (often called the *console, command prompt*, or *terminal*). This is useful in situations where you want to run SQL statements against a MySQL database but don't need a graphical user interface like MySQL Workbench.

At the command line of your computer, enter **mysql** to start the MySQL command line client tool, and supply additional information like:

```
mysql --host localhost --database investment --user rick --password=icu2
```

You can also use single-letter options with a single dash—for example, -h instead of --host; -D for --database; and -u and -p for --user and --password=, respectively.

You specify the host where the MySQL server is located with --host. In this example, the MySQL server is installed on my computer, so I've supplied the value localhost. If you're connecting to a server that's installed on another computer, you can the specify the host, like --host www.nostarch.com, or supply an IP address.

Then, enter the name of the database you want to connect to after --database, your MySQL user ID after --user, and your MySQL password after the --password= option.

NOTE *If your computer doesn't recognize the* mysql *command, try adding the directory where* mysql *is located to your path. If you have a Windows computer, you can add the directory to your* PATH *environment variable using the Environment Variables system properties dialog. If you have a Mac, you usually change the* $PATH *variable in the* .bash_profile *file. You can find more information by searching for "Customizing the Path for MySQL Tools" in the online reference manual at* https://dev.mysql.com/doc/refman/8.0/en/.

You should see this warning:

```
[Warning] Using a password on the command line interface can be insecure.
```

That's because you supplied the database password in plaintext. This isn't a great idea, as anyone looking over your shoulder could see your password. A more secure approach is to let mysql prompt you for it. If you use -p at the command line without specifying the password, the tool will prompt you to enter the password. As you type the letters of the password, asterisks will appear:

```
mysql -h localhost -D investment -u rick -p
Enter password: ****
```

Another approach is to use the MySQL configuration utility to securely store your credentials:

```
> mysql_config_editor set --host=localhost --user=investment --password
Enter password: ****
```

You specify host and user with the --host and --user options. The --password option allows you to enter your password.

Once you have saved credentials, you can use the print --all option to show them:

```
mysql_config_editor print --all
```

The password appears as asterisks:

```
[client]
user = "investment"
password = ****
host = "localhost"
```

Now you can enter the MySQL command line client, mysql, at the command line without having to enter your username, password, or host:

```
mysql -D investment
```

In other words, you can log in to MySQL by providing only the name of the database.

You might wonder why you would use a text-based tool like mysql when more sophisticated graphical tools like MySQL Workbench are available. The mysql tool is particularly useful when you want to run SQL statements that are in a script file. A *script file* is a set of SQL commands saved in a file on your computer. For example, you could create a file called *max_and_min_indexes.sql* that contains the following SQL statements, which get the market index with the smallest and largest values:

```
use investment;

select *
from   market_index
where  market_value =
(
  select  min(market_value)
  from    market_index
);

select *
from   market_index
where  market_value =
(
  select  max(market_value)
  from    market_index
);
```

You can then run the SQL script from your command line using mysql:

```
mysql -h localhost -D investment -u rick -picu2 < min_and_max.sql > min_and_max.txt
```

You used < so that `mysql` will take its input from the *min_and_max.sql* script, and > so that it will write its output to the *min_and_max.txt* file. If you supply the password, in this case `icu2`, don't add a space after `-p`. Strangely, `-picu2` works but `-p icu2` does not.

After you run the command, the output file *min_and_max.txt* should look like this:

```
market_index    market_value
S&P             500 4351.77
market_index                    market_value
Dow Jones Industrial Average    34150.66
```

The `mysql` tool writes a tab between the columns in the file.

NOTE *To see a complete list of options, type* **`mysql --help`** *at the command line.*

Loading Data from a File

Oftentimes you'll get data in the form of files, such as accepting a data feed from another organization. The `load data` command reads data from a file and writes it into a table.

To test loading data from a file into a table, I created a data file on my computer called *market_indexes.txt* in the *C:\Users\rick\market* directory. The file looks like this:

```
Dow Jones Industrial Average   34150.66
Nasdaq                         13552.93
S&P 500                         4351.77
```

The file contains the names and current value of three financial market indexes. It is *tab-delimited*, which means that the fields in the file are separated by the tab character.

In MySQL, load the file into a table like so:

```
use investment;

load data local
infile 'C:/Users/rick/market/market_indexes.txt'
into table market_index;
```

You use the `load data` command and specify `local`, which tells MySQL to look for the data file on your local computer, not on the server where MySQL is installed. By default, `load data` loads tab-delimited files.

After the `infile` keyword, you give the name of the input file you want to load. In this example, you're using the path of a file on a Windows computer. To specify the directory where the file is located on Windows, use forward slashes, as backslashes will result in an error. To load a file in a Mac or Linux environment, use forward slashes as usual.

Take a look at the data that was loaded into the table:

```
select * from market_index;
```

The result is:

```
market_index                  market_value
----------------------------  ------------
Dow Jones Industrial Average     34150.66
Nasdaq                           13552.93
S&P 500                           4351.77
```

There were two fields in the file and two columns in the table, so the fields on the left were loaded into the first column and the fields on the right were loaded into the second column in the table.

Another common data file format is a *comma-separated values (CSV)* file. You could have loaded a data file called *market_indexes.csv* that looks like this:

```
Dow Jones Industrial Average, 34150.66
Nasdaq, 13552.93
S&P 500, 4351.77
```

To load this file, add the syntax `fields terminated by ","` to declare the delimiter in this file as a comma. MySQL uses the commas in the data file to identify the beginning and end of the fields.

```
load data local
infile 'C:/Users/rick/market/market_indexes.csv'
into table market_index
fields terminated by ",";
```

Occasionally, you'll want to load a data file that has a header row, like this:

```
Financial Index, Current Value
Dow Jones Industrial Average, 34150.66
Nasdaq, 13552.93
S&P 500, 4351.77
```

You can have `load data` skip the header by using the `ignore` keyword:

```
load data local
infile 'C:/Users/rick/market/market_indexes.csv'
into table market_index
fields terminated by ","
ignore 1 lines;
```

There was one header line in the data file, so you used the `ignore 1 lines` syntax to prevent the first line from loading into the table. The three rows of data are loaded, but the `Financial Index` and `Current Value` headings in the data file are ignored.

Loading Data to a File

You can provide data to another department or organization by sending data files. One way to write data from the database to a file is to use the syntax `select...into outfile`. You can run queries and select the results to a file rather than to your screen.

You can specify which delimiters you want to use to format the output. Create a CSV file containing the values in the `market_index` table like so:

```
select * from market_index
into outfile 'C:/ProgramData/MySQL/MySQL Server 8.0/Uploads/market_index.csv'
fields terminated by ',' optionally enclosed by '"';
```

You select all values from the `market_index` table and write them to the *market_index.csv* file in the *C:/ProgramData/MySQL/MySQL Server 8.0/Uploads* directory on the host computer.

You use commas as the delimiter in your output file by using the syntax `fields terminated by ','`.

The `optionally enclosed by '"'` line tells MySQL to wrap fields in quotes for any columns that have a `string` data type.

Your *market_index.csv* gets created like this:

```
"Dow Jones Industrial Average",34150.66
"Nasdaq",13552.93
"S&P 500",4351.77
```

The `select...into outfile` syntax can create a file only on the server where MySQL is running. It can't create a file on your local computer.

HOW TO ENABLE LOADING DATA

Depending on how your MySQL environment is configured, running the `load data local` command may produce an error like this:

```
Error Code: 3948. Loading local data is disabled; this must be enabled on
both the client and server sides
```

You or your database administrator can configure MySQL to allow loading local files by setting the `local_infile` system variable to `ON`. You can see the current value of your `local_infile` setting using this command:

```
show global variables like 'local_infile';
```

The result may show that the value is set to `OFF`:

```
Variable_name   Value
-------------   -----
local_infile    OFF
```

If it is set to `OFF`, you won't be able to load files from your client computer. You can set it to `ON` using this command:

```
set global local_infile = on;
```

If you work in a highly secure environment, your database administrator might not want to allow local files to be loaded, and may choose not to change this setting to `ON`.

To load files that reside on the host—the server where MySQL is installed—check for a setting called `secure_file_priv` that controls which directory on the server you can load files from. You can check this value using this command:

```
show global variables like 'secure_file_priv';
```

If the `Value` returned is a directory, then that is the only directory on the server you'll be allowed to import or export files to and from.

```
Variable_name     Value
----------------  ------------------------------------------------
secure_file_priv  C:\ProgramData\MySQL\MySQL Server 8.0\Uploads\
```

In this case, when you use `load data`, the files must be loaded from the *C:\ProgramData\MySQL\MySQL Server 8.0\Uploads* directory on the server. This setting also affects writing to a file with `select...into outfile`. You won't be able to write to files in any other directories on the server.

If the `secure_file_priv` is set to `null`, you won't be able to read from or write to files on the server at all. If it is set to blank (`''`), then you can read from or write to files in any directory on the server.

MySQL Shell

While the MySQL command line client (`mysql`) is a tried-and-true way to run SQL commands that has been used for decades, MySQL Shell (`mysqlsh`) is a newer MySQL command line client tool that can run SQL, Python, or JavaScript commands.

You saw earlier that the `mysql` syntax to run a script called *min_and_max.sql* is:

```
mysql -h localhost -D investment -u rick -picu2 < min_and_max.sql > min_and_max.txt
```

If you prefer, you could use MySQL Shell to run that same script using this command:

```
mysqlsh --sql -h localhost -D investment -u rick -picu2 < min_and_max.sql > min_and_max.txt
```

The syntax is similar, except you call mysqlsh instead of mysql. Also, since mysqlsh can run in SQL, Python, or JavaScript mode, you need to specify --sql to run in SQL mode. (The default mode is JavaScript.)

MySQL Shell comes with a handy utility called *parallel table import* (import-table) that can load large data files to tables faster than load data.

```
mysqlsh ❶ --mysql -h localhost -u rick -picu2 ❷ -- util import-table c:\Users
\rick\market_indexes.txt --schema=investment --table=market_index
```

When you use the import-table utility, you need to call mysqlsh with the --mysql syntax ❶ to use a classic MySQL protocol connection to communicate between the client and the MySQL server.

To run the parallel table import utility, use the -- util syntax and then give the name of the utility you want to use—in this case, import-table ❷. You provide the name of the file you want to load, *c:\Users\rick\market_indexes.txt*, and the database, investment, as well as the table that you want to load the data into, market_index.

The choice to use mysql or mysqlsh is yours. As mysqlsh matures, more developers will move to it and away from mysql. If you have a large data load that is slow to run, using mysqlsh with its parallel table import utility will be considerably faster than using load data.

You can learn more about MySQL Shell at *https://dev.mysql.com/doc/mysql-shell/8.0/en/*.

Summary

In this chapter, you looked at some tips and tricks, including how to avoid making common mistakes, use transactions, support existing systems, and load data to and from files.

In the next chapter, you'll call MySQL from programming languages like PHP, Python, and Java.

15

CALLING MYSQL FROM PROGRAMMING LANGUAGES

In this chapter, you'll write computer programs that use MySQL, focusing on three open source programming languages: PHP, Python, and Java. You'll write programs in each language to select from a table, insert a row into a table, and call a stored procedure.

Regardless of the programming language, you follow the same general steps to call MySQL. First, you establish a connection to the MySQL database using your MySQL database credentials, including the hostname of the MySQL server, the database, user ID, and password. Then, you use that connection to run your SQL statements against the database.

You embed SQL statements in your program and when the program is run, the SQL is executed against the database. If you need to send parameter values to a SQL statement, you use a *prepared statement*, a reusable SQL statement that uses placeholders to temporarily represent the parameters. Then you bind parameter values to the prepared statement, replacing the placeholders with actual values.

If you're retrieving data from the database, you iterate through the results and perform some action, like displaying the results. When you're done, you close the connection to MySQL.

Let's look at some examples using PHP, Python, and Java.

PHP

PHP (a recursive acronym for PHP: Hypertext Preprocessor) is an open source programming language used mostly for web development. Millions of websites have been built with PHP.

NOTE *Instructions for installing PHP are available at* https://www.php.net/manual/en/install.php.

PHP is commonly used with MySQL. Both are part of the *LAMP stack*, a popular software development architecture consisting of Linux, Apache, MySQL, and PHP. (The *P* can also refer to the Python programming language and, less frequently, to Perl.) Many sites use Linux as the operating system; Apache as the web server to receive requests and send back responses; MySQL as the relational database management system; and PHP as the programming language.

To use MySQL from within PHP, you need a PHP *extension*, which enables you to use functionality in your PHP program that isn't included in the core language. Since not all PHP applications need to access MySQL, this functionality is provided as an extension you can load. There are two choices for extensions: *PHP Data Objects (PDO)* and *MySQLi*. You'll specify which you want to load by listing them in your *php.ini* configuration file, like so:

```
extension=pdo_mysql
extension=mysqli
```

The PDO and MySQLi extensions provide different ways to create database connections and execute SQL statements from within your PHP programs. These are object-oriented extensions. (MySQLi is also available as a procedural extension; you'll learn what this means in the section "Procedural MySQLi" later in the chapter.) *Object-oriented programming (OOP)* relies on objects that contain data and can execute code in the form of methods. A *method* is the equivalent of a function in procedural programming; it is a set of instructions you can call to take some action, like running a query or executing a stored procedure.

To use PHP's object-oriented MySQL extensions, you'll create a new `PDO` or `MySQLi` object in your PHP code and use the `->` symbol to call the object's methods.

Let's take a look at each of these extensions, starting with PDO.

PDO

The PDO extension can be used with many relational database management systems, including MySQL, Oracle, Microsoft SQL Server, and PostgreSQL.

Selecting from a Table

In Listing 15-1, you write a PHP program called *display_mountains_pdo.php* that uses PDO to select from a MySQL table called `mountain` in the `topography` database.

```
<?php

❶ $conn = new PDO(
      'mysql:host=localhost;dbname=topography',
      'top_app',
      'pQ3fgR5u5'
  );

❷ $sql = 'select mountain_name, location, height from mountain';

❸ $stmt = $conn->query($sql);

  while ($row = $stmt->fetch(❹ PDO::FETCH_ASSOC)) {
    ❺ echo(
          $row['mountain_name'] . ' | ' .
          $row['location'] . ' | ' .
          $row['height'] . '<br />'
      );
  }
  $conn = ❻ null;
  ?>
```

Listing 15-1: Using PDO to display data from the `mountain` table (display_mountains_pdo.php)

The program begins with the opening tag `<?php` and ends with the closing tag `?>`. The tags tell the web server to interpret the code between them as PHP.

To use MySQL from within PHP, you need to create a connection to your MySQL database by creating a new `PDO` object ❶ and passing in your database credentials. In this case, your hostname is `localhost`, your database name is `topography`, the database user ID is `top_app`, and the password for your MySQL database is `pQ3fgR5u5`.

You can also specify the port by adding it to the end of the line with your host and database name, like so:

```
'mysql:host=localhost;dbname=topography;port=3306',
```

If you don't provide a port, it defaults to `3306`, which is the port normally used to connect to a MySQL server. If your MySQL server instance was configured to run on another port, ask your database administrator to check the configuration files for the port number.

You save the connection as a variable named `$conn`. PHP variables are preceded by a dollar sign. This variable now represents the connection between your PHP program and your MySQL database.

Next, you create a PHP variable called `$sql` that holds your SQL statement ❷.

You call PDO's `query()` method and send it the SQL statement you want to run. In object-oriented programming, the `->` symbol is often used to call an object's instance methods, like `$conn->query()`. You save the statement and its results as an array variable called `$stmt` ❸.

NOTE *In contrast to the* `->` *symbol used to access an object's* instance *methods and variables, the double-colon syntax* (`::`) *is used to access an object's* static *members, like any constant variables. It is known as the* scope resolution *operator (also sometimes called* paamayim nekudotayim, *which is Hebrew for "double colon").*

An *array* is a type of variable you can use to store a group of values. It uses an *index* to identify one value in the group. You use PDO's `fetch()` method to fetch each row from `$stmt` using a *mode*, which controls how the data gets returned to you. Here, the mode `PDO::FETCH_ASSOC` ❹ returns an array that is indexed by the database table's column names, like `$row['mountain_name']`, `$row['location']`, and `$row['height']`. If you had used the mode `PDO::FETCH_NUM`, it would have returned an array that is indexed by the column number starting at zero, like `$row[0]`, `$row[1]`, and `$row[2]`. Other modes can be found in PHP's online manual at *https://php.net.*

Next, the `while` loop will loop through each row that was fetched. You use the `echo()` command ❺ to display each column separated by the pipe (`|`) character. The `<br />` HTML tag at the end of your `echo()` statement will create a line break after each line in your browser.

Finally, you close the connection by setting it to `null` ❻.

Navigate to *http://localhost/display_mountains_pdo.php* to see the results of your PHP program, shown in Figure 15-1.

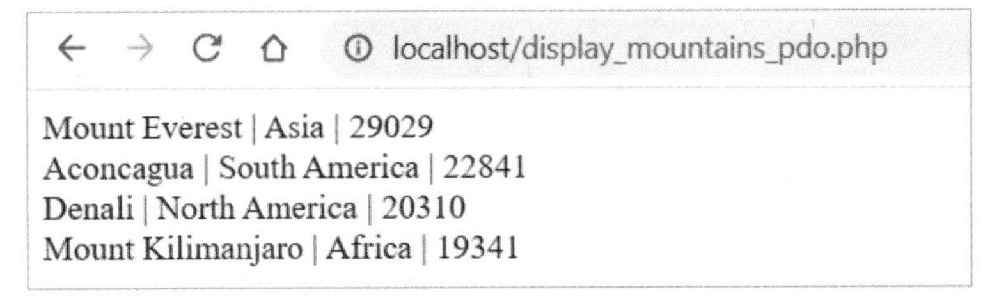

Figure 15-1: The results of display_mountains_pdo.php

You've successfully accessed MySQL via PDO to select data from the `mountain` table and return each column separated by the pipe character.

Inserting a Row into a Table

Now you'll create a new PHP program called *add_mountain_pdo.php* that inserts a new row in the `mountain` table using PDO.

In Listing 15-2, you'll use a prepared statement, which, as mentioned earlier, uses placeholders to represent values in a SQL statement. Then, you'll replace those placeholders with actual values from PHP variables. Using prepared statements is a good security practice because it helps protect against SQL injection attacks, which are a common way for hackers to run malicious SQL code against your database.

```
<?php

❶ $conn = new PDO(
      'mysql:host=localhost;dbname=topography',
      'top_app',
      'pQ3fgR5u5'
  );

  $new_mountain = 'K2';
  $new_location = 'Asia';
  $new_height = 28252;

  $stmt = $conn->❷prepare(
      'insert into mountain (mountain_name, location, height)
       values (❸:mountain, :location, :height)'
  );

❹ $stmt->bindParam(':mountain', $new_mountain, PDO::PARAM_STR);
  $stmt->bindParam(':location', $new_location, PDO::PARAM_STR);
  $stmt->bindParam(':height',   $new_height,   PDO::PARAM_INT);

  $stmt->❺execute();

  $conn = null;
  ?>
```

Listing 15-2: Using PDO to insert a row into the `mountain` *table (*`add_mountain_pdo.php`*)*

As in Listing 15-1, you first create a connection to your MySQL database ❶. You have three PHP variables called `$new_mountain`, `$new_location`, and `$new_height` that hold the name, location, and height of the mountain, respectively, that you want to insert into your `mountain` table.

You use the connection's `prepare()` method ❷ to create a prepared statement that uses named placeholders for your values. You write the `insert` SQL statement, but instead of including the actual values you want to insert, you use placeholders ❸. Your named placeholders are `:mountain`, `:location`, and `:height`. Named placeholders are preceded by a colon.

NOTE *PDO also allows you to use question marks instead of named placeholders, like so:*

```
$stmt = $conn->prepare(
    'insert into mountain (mountain_name, location, height)
     values (?, ?, ?)'
);
```

I recommend using named placeholders instead of question marks, however, because they make your code more readable.

Next, you replace the placeholders with actual values using the `bindParam()` method ❹, which links, or binds, a placeholder with a variable. You bind the first placeholder to the `$new_mountain` variable, which replaces `:mountain` with the value `K2`. You bind the second placeholder to the `$new_location` variable, replacing `:location` with the value `Asia`. You bind the third placeholder to the `$new_height` variable, replacing `:height` with the value `28252`.

Then, you specify the type of data the variables represent. The mountain and location are strings, so you use `PDO::PARAM_STR`. The height is an integer, so you use `PDO::PARAM_INT`.

When you call the statement's `execute()` method ❺, your statement is executed and your new row gets inserted into the `mountain` table.

Calling a Stored Procedure

Next, you'll write a PHP program named *find_mountains_by_loc_pdo.php* that calls a MySQL stored procedure, `p_get_mountain_by_loc()`.

You'll provide the stored procedure with a parameter for the location you want to search for; in this example, you'll search for mountains in `Asia`. Your PHP program will call the stored procedure and return the number of mountains in the `mountain` table that are in Asia (see Listing 15-3).

```
<?php

$conn = new PDO(
    'mysql:host=localhost;dbname=topography',
    'top_app',
    'pQ3fgR5u5'
);

$location = 'Asia';

$stmt = $conn->prepare('❶call p_get_mountain_by_loc(❷:location)');
$stmt->❸bindParam(':location', $location, PDO::PARAM_STR);

$stmt->❹execute();

❺ while ($row = $stmt->fetch(PDO::FETCH_ASSOC)) {
    echo(
        $row['mountain_name'] . ' | ' .
        $row['height'] . '<br />'
    );
}
$conn = null;
?>
```

Listing 15-3: Using PDO to call a stored MySQL procedure (find_mountains_by_loc_pdo.php)

You use the `call` statement ❶ in your prepared statement to call the stored procedure. Then you create a named placeholder, `:location` ❷, and use `bindParam` ❸ to replace `:location` with the value in the `$location` variable, which evaluates to `Asia`.

Next, you execute the stored procedure ❹. Then you use a `while` statement ❺ to select the rows returned from the stored procedure. You display them to the user using the `echo` command. Finally, you end the connection. The results are shown in Figure 15-2.

Figure 15-2: The results of find_mountains_by_loc_pdo.php

You can add more functionality to these programs. For example, you might choose to allow the user to select the location they want to see, rather than hardcoding `Asia` in the PHP program. You could even check for errors connecting to the database or calling your stored procedure, and display detailed error messages to the user when there is a problem.

HARDCODING CREDENTIALS

Regardless of the programming language you use, you shouldn't hardcode database credentials.

While it's fine to use code that includes your host, database name, user ID, and password for demonstration purposes,

```
$conn = new PDO(
    'mysql:host=localhost;dbname=topography',
    'top_app',
    'pQ3fgR5u5'
);
```

in a real application, you wouldn't display such sensitive information in plaintext. Instead, the database connection information is often stored and loaded from a configuration file that has its file permissions set to control who can access the information.

Object-Oriented MySQLi

The MySQL Improved (MySQLi) extension is the upgraded version of an old legacy PHP extension that was called MySQL. In this section, you'll learn how to use the object-oriented version of MySQLi.

Selecting from a Table

In Listing 15-4, you write a PHP program using the object-oriented MySQLi to select from your `mountain` table.

```
<?php

$conn = ❶ new mysqli(
          'localhost',
          'top_app',
          'pQ3fgR5u5',
          'topography'
);

$sql = 'select mountain_name, location, height from mountain';

$result = $conn->❷query($sql);

while ($row = ❸ $result->fetch_assoc()) {
  echo(
    $row['mountain_name'] . ' | ' .
    $row['location'] . ' | ' .
    $row['height'] . '<br />'
  );
}
❹ $conn->close();
?>
```

Listing 15-4: Using object-oriented MySQLi to display data from the `mountain` *table (*display_mountains_mysqli_oo.php*)*

You establish a connection to MySQL by creating a `mysqli` object ❶ and passing in the host, user ID, password, and database. You run your query using the connection's `query()` method ❷ and save the results to a PHP variable called `$result`.

You iterate through the resulting rows, calling the `$result fetch_assoc()` method ❸ so that you can reference the columns as indexes, like `$row ['mountain_name']`. Then you print the values of those columns and close your connection with the `close()` method ❹.

The result is displayed in Figure 15-3.

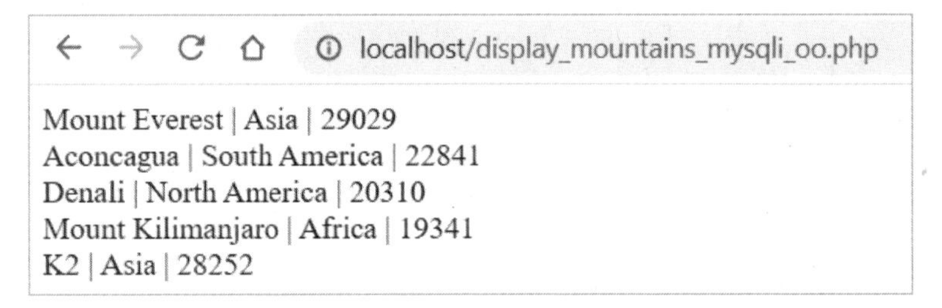

Figure 15-3: The results of display_mountains_mysqli_oo.php

Inserting a Row into a Table

Now, you'll create a PHP program to insert a row into the `mountain` table using object-oriented MySQLi (see Listing 15-5).

```
<?php

$conn = new mysqli(
          'localhost',
          'top_app',
          'pQ3fgR5u5',
          'topography'
);

$new_mountain = 'Makalu';
$new_location = 'Asia';
$new_height = 27766;

❶ $stmt = $conn->prepare(
     'insert into mountain (mountain_name, location, height)
      values (?, ?, ?)'
);

❷ $stmt->bind_param('ssi',$new_mountain,$new_location,$new_height);
$stmt->execute();
$conn->close();
?>
```

Listing 15-5: Using object-oriented MySQLi to insert a row into the `mountain` *table (add_mountain_mysqli_oo.php)*

Once you've established your connection, you use a prepared statement with question marks as your placeholders ❶. Then, you replace your question mark placeholders with values using the `bind_param()` method ❷.

NOTE *While PDO allows you to use named placeholders, MySQLi requires you to use question marks as placeholders.*

With MySQLi, you can provide the data types of the bind variables as a string. The first parameter you send to `bind_param()` is the value `ssi`, which indicates that you want to replace the first and second placeholders with a string (`s`) value, and the third placeholder with an integer (`i`) value. You can also choose `d` if the bind variable has a data type of `double` (a double-precision floating-point number) or `b` if the bind variable has a data type of `blob` (a binary large object).

Finally, you execute the prepared statement with `execute()` and close your connection. When you run the program, it inserts a new mountain—`Makalu`—into your `mountain` table.

Calling a Stored Procedure

Listing 15-6 shows a PHP program that uses object-oriented MySQLi to execute a stored procedure.

```
<?php

$conn = new mysqli(
          'localhost',
```

```
        'top_app',
        'pQ3fgR5u5',
        'topography'
);

$location = 'Asia';

$stmt = $conn->prepare('call p_get_mountain_by_loc(?)');
❶ $stmt->bind_param('s', $location);
$stmt->execute();

$result = $stmt->get_result();

while ($row = $result->fetch_assoc()) {
  echo(
    $row['mountain_name'] . ' | ' .
    $row['height'] . '<br />'
  );
}

$conn->close();
?>
```

*Listing 15-6: Using object-oriented MySQLi to call a stored MySQL procedure (*find_mountains_by_loc_mysqli_oo.php*)*

You use a prepared statement that calls the `p_get_mountain_by_loc()` stored procedure. It has one question mark placeholder that represents the location of the mountains you want to search for. You bind the location, replacing the `?` with `Asia`. You send `s` as the first parameter to the `bind_param()` method to indicate that the location is a string ❶.

Once you execute the statement and loop through your results, the name and height of the Asian mountains in your table are displayed.

The results are shown in Figure 15-4.

Figure 15-4: The results of find_mountains_by_loc_mysqli_oo.php

Procedural MySQLi

MySQLi is also available as a procedural extension. The procedural version of MySQLi looks similar to the object-oriented version, but instead of using `->` syntax to call methods, like `$conn->close()`, you'll use functions that start with the text `mysqli_`, like `mysqli_connect()`, `mysqli_query()`, and `mysqli_close()`.

Procedural programming treats data and procedures as two different entities. It uses a top-down approach where you write code giving instructions in order from beginning to end, and call procedures—or functions—that contain code to handle specific tasks.

> **SHOULD I USE MYSQLI OR PDO?**
>
> When working with an existing PHP application, continue using the extension that's already being used in the codebase. But which extension should you use if you're creating a new system?
>
> If you want to migrate your application from MySQL to another database system like Oracle, PostgreSQL, or Microsoft SQL Server, PDO is a good choice. MySQLi works only with MySQL, whereas PDO also works with other database systems, which makes migrating your application to another database system easier. If you support different PHP applications that use different database systems, you can learn to use PDO with MySQL and use the same approach for those other systems as well.
>
> PDO also allows you to use named placeholders instead of question marks, making your code a bit clearer.

Selecting from a Table

In Listing 15-7, you write a PHP program to select from your `mountain` table using the procedural version of MySQLi.

```
<?php

$conn = mysqli_connect(
          'localhost',
          'top_app',
          'pQ3fgR5u5',
          'topography'
);

$sql = 'select mountain_name, location, height from mountain';

$result = mysqli_query($conn, $sql);

while ($row = mysqli_fetch_assoc($result)) {
  echo(
    $row['mountain_name'] . ' | ' .
    $row['location'] . ' | ' .
    $row['height'] . '<br />'
  );
}
mysqli_close($conn);
?>
```

Listing 15-7: Using procedural MySQLi to display data from the `mountain` *table (*display_mountains_mysqli_procedural.php*)*

You use MySQLi's `mysqli_connect()` function to connect to the database with your database credentials. You define a variable called `$sql` that holds your SQL statement. Next, you use MySQLi's `mysqli_query()` function to run the query using your connection, and save the results to the `$result` variable.

Then, you fetch the results using the `mysql_fetch_assoc()` function so you can reference the resulting `$row` variables using indexes matching the database column names, like `$row['mountain_name']`.

You print the results using the `echo` command and add a pipe (`|`) delimiter between the values. The HTML `<br />` tag will add a line break after each row in your browser.

Finally, you close the connection using the `mysqli_close()` function.

The results are displayed in Figure 15-5.

```
localhost/display_mountains_mysqli_procedural.php

Mount Everest | Asia | 29029
Aconcagua | South America | 22841
Denali | North America | 20310
Mount Kilimanjaro | Africa | 19341
K2 | Asia | 28252
Makalu | Asia | 27766
```

Figure 15-5: The results of display_mountains_mysqli_procedural.php

Inserting a Row into a Table

Now, you'll create a PHP program to insert a row into your `mountain` table using procedural MySQLi (Listing 15-8).

```
<?php

$conn = mysqli_connect(
          'localhost',
          'top_app',
          'pQ3fgR5u5',
          'topography'
);

$new_mountain = 'Lhotse';
$new_location = 'Asia';
$new_height = 27940;

$stmt = mysqli_prepare(
    $conn,
    'insert into mountain (mountain_name, location, height)
     values (?, ?, ?)'
);

mysqli_stmt_bind_param(
    $stmt,
    'ssi',
    $new_mountain,
```

```
    $new_location,
    $new_height
);

mysqli_stmt_execute($stmt);
mysqli_close($conn);
?>
```

Listing 15-8: Using procedural MySQLi to insert a row into the `mountain` *table (*add_mountain_mysqli_procedural.php*)*

The program inserts a new mountain called `Lhotse` into your `mountain` table. The program's logic is similar to the programs you've seen before: you create a connection using your database credentials, use a prepared statement with ? placeholders, bind values to replace the placeholders, execute the statement, and close the connection.

Calling a Stored Procedure

The PHP code to execute a stored procedure using procedural MySQLi is shown in Listing 15-9.

```
<?php

$conn = mysqli_connect(
          'localhost',
          'top_app',
          'pQ3fgR5u5',
          'topography'
);

$location = 'Asia';

$stmt = mysqli_prepare($conn, 'call p_get_mountain_by_loc(?)');
mysqli_stmt_bind_param($stmt, 's', $location);
mysqli_stmt_execute($stmt);
$result = mysqli_stmt_get_result($stmt);

while ($row = mysqli_fetch_assoc($result)) {
  echo(
    $row['mountain_name'] . ' | ' .
    $row['height'] . '<br />'
  );
}
mysqli_close($conn);
?>
```

*Listing 15-9: Using procedural MySQLi to call a stored MySQL procedure (*find_mountains_by_loc_mysqli_procedural.php*)*

You use a prepared statement to call the procedure and a question mark placeholder to represent the stored procedure's parameter. You bind the `$location` PHP variable and specify `s` (string) as the data type. Then, you execute the statement and fetch and iterate through the resulting rows, printing the mountain name and height for each row in your `mountain` table that is in Asia. Finally, you close your connection.

The results are shown in Figure 15-6.

Figure 15-6: The results of find_mountains_by_loc_mysqli_procedural.php

Python

Python is an open source programming language with concise and readable syntax. It's worthwhile to learn Python because it can be used for many different types of programming—from data science and math to video games, web development, and even artificial intelligence!

NOTE *Instructions for installing Python can be found at Python.org at* https://wiki.python.org/moin/BeginnersGuide/Download/.

Python's syntax is unique in that it places a lot of importance on indentation. Other languages use curly brackets to group a block of code, as in this PHP code:

```
if ($temp > 70) {
    echo "It's hot in here. Turning down the temperature.";
    $new_temp = $temp - 2;
    setTemp($new_temp);
}
```

Because your block of PHP code starts with { and ends with }, the indentation of the lines of code within the block doesn't matter; it's just for readability. The following code runs just as well in PHP:

```
if ($temp > 70) {
    echo "It's hot in here. Turning down the temperature.";
$new_temp = $temp - 2;
setTemp($new_temp);
}
```

Python, on the other hand, doesn't use curly brackets to identify blocks of code. It relies on indentation:

```
if temp > 70:
    print("It's hot in here. Turning down the temperature.");
    new_temp = temp - 2
    set_temp(new_temp)
```

If the temperature is over 70 degrees, this example will print `It's hot in here` and will turn down the temperature 2 degrees.

But if you change the indentation in Python, the program does something different:

```
if temp > 70:
    print("It's hot in here. Turning down the temperature.")
new_temp = temp - 2
set_temp(new_temp)
```

The message `It's hot in here` will still print only when the temperature is more than 70, but now the temperature will be turned down 2 degrees regardless. That's probably not what you intended.

NOTE *To access MySQL from your Python code, you can use* MySQL Connector/Python, *a driver that allows Python to communicate with MySQL.*

Selecting from a Table

In Listing 15-10, you write a Python program called *display_mountains.py* to select from the `mountain` table and display the results.

```
import mysql.connector

❶ conn = mysql.connector.connect(
    user='top_app',
    password='pQ3fgR5u5',
    host='localhost',
    database='topography')

❷ cursor = conn.cursor()

cursor.execute('select mountain_name, location, height from mountain')

❸ for (mountain, location, height) in cursor:
    print(mountain, location, height)

conn.close()
```

Listing 15-10: Using Python to display data from the `mountain` *table (*display_mountains.py*)*

In the first line of your code, you import MySQL Connector/Python with `mysql.connector`. Then you create a connection to your MySQL database ❶ by calling the `connect()` method with your database credentials. You save this connection as a Python variable called `conn`.

You use the connection to create a cursor that you save as a variable called `cursor` ❷. Next, you use the `cursor` `execute()` method to run a SQL query that selects from the `mountain` table. A `for` loop is one type of loop that allows you to loop, or iterate, through values. Here, you use a `for` loop ❸ to iterate through the rows in the cursor, printing the mountain name, location, and height of each mountain as you go. The looping will continue until there are no more rows to loop through in `cursor`.

Lastly, you close the connection with conn.close().

You can navigate to your operating system's command prompt and run the Python program to see the results:

```
> python display_mountains.py
Mount Everest Asia 29029
Aconcagua South America 22841
Denali North America 20310
Mount Kilimanjaro Africa 19341
K2 Asia 28252
Makalu Asia 27766
Lhotse Asia 27940
```

Your Python program selected all the rows from your mountain table and displayed the data from the table.

While your database credentials are included in your Python program in this example, you'd typically put sensitive information in a Python file called *config.py* to separate them from the rest of your code.

Inserting a Row into a Table

Now, you'll write a Python program called *add_mountain.py* to insert a row into the mountain table (Listing 15-11).

```
import mysql.connector

conn = mysql.connector.connect(
    user='top_app',
    password='pQ3fgR5u5',
    host='localhost',
    database='topography')

cursor = conn.cursor(prepared=True)
❶ sql = "insert into mountain(mountain_name, location, height) values (?,?,?)"
❷ val = ("Ojos Del Salado", "South America", 22615)
cursor.execute(sql, val)
❸ conn.commit()
cursor.close()
```

Listing 15-11: Using Python to insert a row into the mountain table (add_mountain.py)

Using your connection, you create a cursor that allows you to use prepared statements.

You create a Python variable called sql that contains the insert statement ❶. Python can use either ? or %s for placeholders in prepared statements. (The letter *s* has nothing to do with the data type or values here; that is, the placeholder %s isn't just for strings.)

You create a variable called val ❷ that contains the values you want to insert into the table. Then you call the cursor execute() method, passing in your sql and val variables. The execute() method binds the variables, replacing the ? placeholders with the values, and executes the SQL statement.

You need to commit the statement to the database by calling the connection `commit()` method ❸. By default, MySQL Connector/Python doesn't automatically commit, so if you forget to call `commit()`, the changes won't be applied to your database.

Calling a Stored Procedure

Listing 15-12 shows a Python program called *find_mountains_by_loc.py* that calls the `p_get_mountain_by_loc()` stored procedure and sends a parameter value of `Asia` to display only the mountains in the table that are in Asia.

```
import mysql.connector

conn = mysql.connector.connect(
    user='top_app',
    password='pQ3fgR5u5',
    host='localhost',
    database='topography')

cursor = conn.cursor()

❶ cursor.callproc('p_get_mountain_by_loc', ['Asia'])

❷ for results in cursor.stored_results():
    for record in results:
        print(record[0], record[1])

conn.close()
```

Listing 15-12: Using Python to call a stored procedure (find_mountains_by_loc.py)

You call the `cursor` `callproc()` method to call your stored procedure, sending it a value of `Asia` ❶. Then, you call the `cursor` `stored_results()` method to get the results of the stored procedure, and you iterate through those results using a `for` loop to get the record for each mountain ❷.

Python uses zero-based indexes, so `record[0]` represents the first column that was returned from the stored procedure for the row—in this example, the mountain name. To print the second column, the mountain's height, you use `record[1]`.

Run the Python program from the command line to see the results:

```
> python find_mountains_by_loc.py
Mount Everest 29029
K2 28252
Makalu 27766
Lhotse 27940
```

Java

Java is an open source, object-oriented programming language that is commonly used for everything from mobile app development to desktop applications to web apps.

NOTE *Instructions for installing Java are available on Java.com at* https://www.java.com/en/download/help/download_options.html.

There are lots of build tools and integrated development environments (IDEs) for Java, but for these examples, you'll work from the command line. Let's go over the basics before we start looking at examples.

You'll create a Java program that ends in the file extension *.java*. To run a Java program, you first compile it to a *.class* file using the `javac` command. This file is in bytecode format. *Bytecode* is a machine-level format that runs in the Java Virtual Machine (JVM). Once a program is compiled, you run it using the `java` command.

Here you create a Java program called *MountainList.java* and compile it to bytecode:

```
javac MountainList.java
```

That command creates a bytecode file called *MountainList.class*. To run it, you use this command:

```
java MountainList
```

NOTE MySQL Connector/J *is a Java Database Connectivity (JDBC) driver, which lets you communicate between Java and MySQL. You can use MySQL Connector/J to connect to MySQL databases, run SQL statements, and process the results. You can also use it to run stored procedures from within Java.*

Selecting from a Table

As with the other programming languages, you'll start by writing a Java program called *MountainList.java* that selects a list of mountains from your MySQL `mountain` table (Listing 15-13).

```
import java.sql.*; ❶

public class MountainList {
public static void main(String args[]) { ❷
    String url = "jdbc:mysql://localhost/topography";
    String username = "top_app";
    String password = "pQ3fgR5u5";

    try { ❸
      Class.forName("com.mysql.cj.jdbc.Driver"); ❹
      Connection ❺ conn = DriverManager.getConnection(url, username, password);
      Statement stmt = conn.createStatement(); ❻
      String sql = "select mountain_name, location, height from mountain";
      ResultSet rs = stmt.executeQuery(sql); ❼
      while (rs.next()) {
        System.out.println(
          rs.getString("mountain_name") + " | " +
          rs.getString("location") + " | " +
          rs.getInt("height");
```

```
        );
      }
      conn.close();
    } catch (Exception ex) {
      System.out.println(ex);
    }
  }
}
```

Listing 15-13: Using Java to display data from the `mountain` table (MountainList.java)

First, you import the `java.sql` package ❶ to give you access to Java objects for using a MySQL database, like `Connection`, `Statement`, and `ResultSet`.

You create a Java class called `MountainList` that has a `main()` method, which is automatically executed when you run the program ❷. In the `main()` method, you create a connection to your MySQL database by providing your database credentials. You save this connection as a Java variable called `conn` ❺.

You load the Java class for MySQL Connector/J, `com.mysql.cj.jdbc.Driver`, using the `Class.forName` command ❹.

Using the `Connection` `createStatement()` method, you create a `Statement` ❻ to execute SQL against the database. The `Statement` returns a `ResultSet` ❼, which you loop through to display the name, location, and height of each mountain in the database table. You close the connection when you're done.

Notice that many of these Java commands are wrapped in a `try` block ❸. This way, if there are problems running these commands, Java will throw an *exception* (or error) that you can catch in your corresponding `catch` statement. In this case, when an exception is thrown, control is passed to the `catch` block and you display the exception to the user.

In Python and PHP, wrapping your code in a `try...catch` block is best practice, but optional. (The syntax in Python is `try/except`.) But in Java, you *must* use a `try...catch` block. If you try to compile the Java code without it, you'll get an error saying that exceptions `must be caught or declared to be thrown`.

Compile and run your Java program from the command line, and see the results:

```
> javac MountainList.java
> java MountainList
Mount Everest | Asia | 29029
Aconcagua | South America | 22841
Denali | North America | 20310
Mount Kilimanjaro | Africa | 19341
K2 | Asia | 28252
Makalu | Asia | 27766
Lhotse | Asia | 27940
Ojos Del Salado | South America | 22615
```

Inserting a Row into a Table

In Listing 15-14, you'll write a Java program to insert a row into the `mountain` table.

```
import java.sql.*;

public class MountainNew {
  public static void main(String args[]) {
    String url = "jdbc:mysql://localhost/topography";
    String username = "top_app";
    String password = "pQ3fgR5u5";

    try {
      Class.forName("com.mysql.cj.jdbc.Driver");
      Connection conn = DriverManager.getConnection(url, username, password);
      String sql = "insert into mountain(mountain_name, location, height) " +
                   "values (?,?,?)";
    ❶ PreparedStatement stmt = conn.prepareStatement(sql);
      stmt.setString(1, "Kangchenjunga");
      stmt.setString(2, "Asia");
      stmt.setInt(3, 28169);
    ❷ stmt.executeUpdate();
      conn.close();
    } catch (Exception ex) {
      System.out.println(ex);
    }
  }
}
```

Listing 15-14: Using Java to insert a row into the `mountain` *table (*`MountainNew.java`*)*

Your SQL statement uses question marks as placeholders. You use a `PreparedStatement` this time ❶ instead of a `Statement` so that you can send parameter values. You bind the parameter values using the `setString()` and `setInt()` methods. Then you call the `executeUpdate()` method ❷, which is used to insert, update, or delete rows in your MySQL table.

Calling a Stored Procedure

Listing 15-15 shows a Java program to execute a MySQL stored procedure.

```
import java.sql.*;

public class MountainAsia {
  public static void main(String args[]) {
    String url = "jdbc:mysql://localhost/topography";
    String username = "top_app";
    String password = "pQ3fgR5u5";

    try {
      Class.forName("com.mysql.cj.jdbc.Driver");
      Connection conn = DriverManager.getConnection(url, username, password);
      String sql = "call p_get_mountain_by_loc(?)";
    ❶ CallableStatement stmt = conn.prepareCall(sql);
      stmt.setString(1, "Asia");
      ResultSet rs = stmt.executeQuery();
```

```
      while (rs.next()) {
        System.out.println(
          rs.getString("mountain_name") + " | " +
          rs.getInt("height")
        );
      }
      conn.close();
    } catch (Exception ex) {
      System.out.println(ex);
    }
  }
}
```

*Listing 15-15: Using Java to call a MySQL stored procedure (*MountainAsia.java*)*

This time, you use a `CallableStatement` ❶ instead of `Statement` to call stored procedures. You set the first (and only) parameter to `Asia` and execute your query using `CallableStatement`'s `executeQuery()` method. Then you iterate through the results, displaying each mountain name and height.

The results are:

```
Mount Everest | 29029
K2 | 28252
Makalu | 27766
Lhotse | 27940
Kangchenjunga | 28169
```

OBJECT-RELATIONAL MAPPING

Object-relational mapping (ORM) tools take a different approach to using MySQL from within programming languages like the ones you've seen in this chapter. ORM allows you to interact with your database using your favorite object-oriented programming language instead of using SQL. ORM makes the data in the database available to you in the form of objects that you can manipulate in your code.

Summary

In this chapter, you looked at calling MySQL from programming languages. You learned that SQL statements are often embedded and run from within programs. You saw that the same database table that can be accessed from MySQL Workbench can also be accessed using PHP, Python, Java, or any number of other tools or languages.

In the next chapter, you'll work on your first project using MySQL: creating a functioning weather database. You'll build scripts to accept a weather feed hourly and load it into your MySQL database.

PART V

PROJECTS

Congratulations! You now know enough about MySQL to begin building meaningful projects. In this part of the book, you'll work through three projects that will teach you new skills and provide a deeper understanding of the material presented so far.

These projects, outlined below, are independent of one another and can be completed in any order:

Building a Weather Database

You'll build a database to store current weather data for a trucking company using technologies including cron, Bash, and SQL scripts.

Tracking Changes to Voter Data with Triggers

You'll build a database to hold election data and build triggers on the tables to track changes to the voter data.

Protecting Salary Data with Views

You'll build a database that holds company data and allows access to the salary data only as needed. You'll hide salaries from most users.

16

BUILDING A WEATHER DATABASE

In this project, you'll build a weather database for a trucking company. The company transports items up and down the East Coast of the United States and needs a way to get the current weather for the major cities its drivers travel to.

The company already has a MySQL database set up that contains trucking data, but you need to add a new database detailing the current weather conditions for areas the truckers drive through. This will allow you to incorporate weather data into the existing trucking application to show the weather's impact on scheduling and warn drivers of hazardous conditions like black ice, snow, and extreme temperatures.

You'll get these weather data files from a third-party company that provides weather data. That company has agreed to send you a CSV file every hour. Recall from Chapter 14 that a CSV file is a text file that contains data and uses a comma as the delimiter between fields.

The company providing the weather data will use FTP (File Transfer Protocol), a standard communication protocol that allows files to be transferred between computers, to deliver the *weather.csv* file to your */home/*

weather_load/ directory on one of your Linux servers. The data file will arrive approximately every hour, but there can be delays, meaning the files may not arrive exactly at hourly intervals. For that reason, you'll write a program that will run every 5 minutes to check for the file and load it into your database when it's available.

Once you've reviewed the necessary technologies, you'll begin your project by creating a new database called `weather` with two tables: `current_weather_load` and `current_weather`. You'll load the data from the file into the `current_weather_load` table. Once you ensure that the data loads without any problems, you'll copy the data from `current_weather_load` to the `current_weather` table, which is the table that your trucking application will use. You can find the *weather.csv* data file at *https://github.com/ricksilva/mysql_cc/tree/main/chapter_16*.

Technologies You'll Use

For this project, you'll use other technologies in addition to MySQL, including cron and Bash. These technologies allow you to schedule the loading of your weather data, check whether the data file is available, and create a logfile containing any load errors.

cron

In order to schedule a script to run every 5 minutes, you'll use *cron*, a scheduler available on Unix-like operating systems (Unix, Linux, and macOS). It is also available on Windows through the Windows Subsystem for Linux (WSL), which lets you run a Linux environment on a Windows computer. To install WSL, enter **`wsl --install`** on the command line.

The tasks that you schedule in cron are called *cron jobs*, and they run in the background, not attached to a terminal. You can schedule jobs by adding them to a configuration file called the *crontab* (*cron table*) file.

> **LEARNING MORE ABOUT CRON**
>
> If you'd like to learn more about cron, type `man cron` at your command prompt. The `man` command is used to display a manual of commands that can be run, and `cron` specifies that you want information about cron, including a list of options you can use with it.
>
> If cron isn't installed on your computer, you'll need to search online for the command to install it in your particular environment.
>
> You can start and stop cron, but the commands to do so vary, so try searching for the command using `man cron`. You should also find the command to get cron's status in order to see if it's started or stopped.

You can get a list of your scheduled cron jobs by typing `crontab -l`. If you need to edit your crontab configuration file, type `crontab -e`. The `-e` option will open a text editor where you can add, modify, or delete jobs from your crontab file.

To schedule a cron job, you must provide six pieces of information, in this order:

1. Minute (0–59)
2. Hour (0–23)
3. Day of the month (1–31)
4. Month (1–12)
5. Day of the week (0–6) (Sunday to Saturday)
6. The command or script to run

For example, if you wanted to schedule a script called *pi_day.sh* to run, you'd type `crontab -e` and add a crontab entry that looks like this:

```
14 3 14 3 * /usr/local/bin/pi_day.sh
```

With this cron job in place, the *pi_day.sh* script in the */usr/local/bin/* directory will execute every year on March 14 at 3:14 AM. Since the day of the week has been set to * (the wildcard), the job will execute on whatever day of the week March 14th happens to be on that year.

Bash

Bash is a shell and command language available in Unix and Linux environments. You could use any number of tools or languages, but I've chosen Bash because of its popularity and relative simplicity. Bash scripts usually have the extension *.sh*, like *pi_day.sh* in the preceding example. In this chapter's project, you'll write a Bash script called *weather.sh* that cron will run every 5 minutes. This script will check if a new data file has arrived and call SQL scripts to load the data into your database if it has.

NOTE *You can learn more about Bash at* https://linuxconfig.org/bash-scripting-tutorial-for-beginners *or in* The Linux Command Line, 2nd edition, *by William Shotts (No Starch Press, 2019).*

SQL Scripts

SQL scripts are text files that contain SQL commands. For this project, you'll write two SQL scripts called *load_weather.sql* and *copy_weather.sql*. The *load_weather.sql* script will load the data from the CSV file into the `current_weather_load` table and alert you to any load issues. The *copy_weather.sql* script will copy the weather data from the `current_weather_load` table to the `current_weather` table.

Project Overview

You'll schedule a cron job to run the *weather.sh* script every 5 minutes. If a new *weather.csv* data file exists, it will be loaded into the `current_weather_load` table. If it is loaded without errors, the data in the `current_weather_load` table will be copied to the `current_weather` table, where it will be used by your application. Figure 16-1 shows the flow of the project.

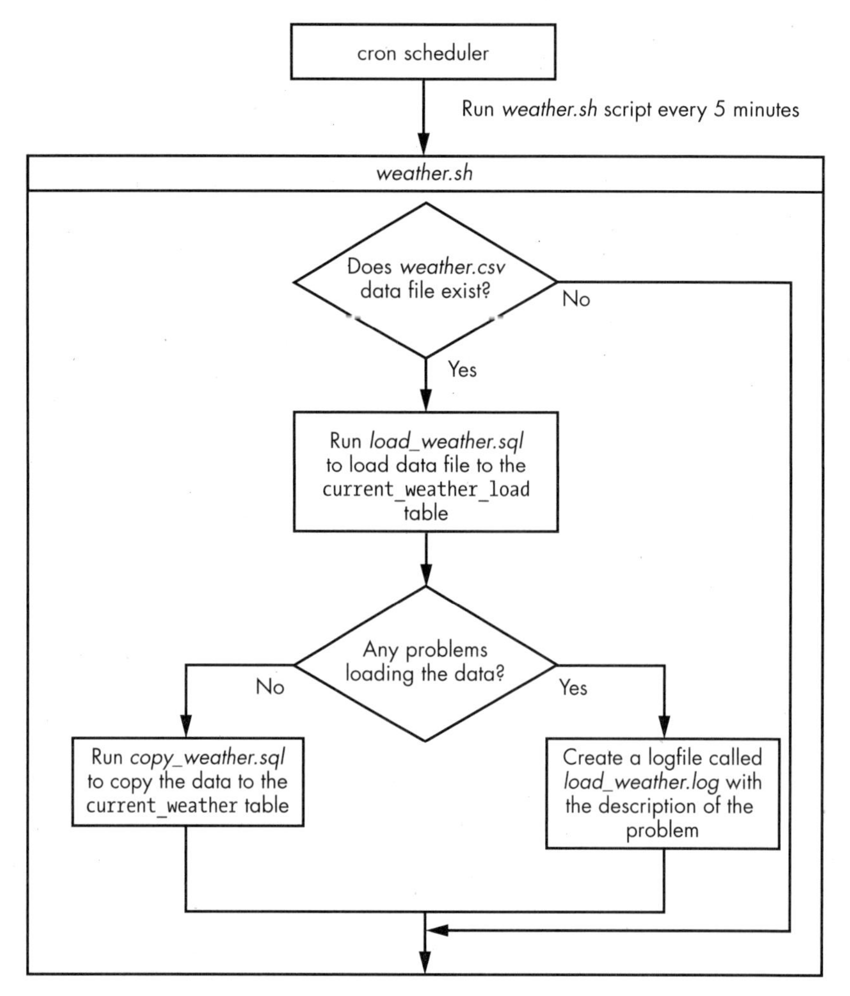

Figure 16-1: An overview of your weather project

If there isn't a new *weather.csv* file available, the *weather.sh* script exits without running the rest of the commands in the Bash script that load the data and log errors. If the file was loaded and there aren't any errors in *load_weather.log*, the Bash script will call *copy_weather.sql* to copy the data you just loaded in the `current_weather_load` table to the `current_weather` table.

The Data File

Since the trucking company travels up and down the US East Coast, you've requested the weather for the following locations:

- Portland, Maine
- Boston, Massachusetts
- Providence, Rhode Island
- New York, New York
- Philadelphia, Pennsylvania
- Washington, DC
- Richmond, Virginia
- Raleigh, North Carolina
- Charleston, South Carolina
- Jacksonville, Florida
- Miami, Florida

The CSV data file will include the fields listed in Table 16-1.

Table 16-1: Fields in the CSV Data File

Field name	Description
`station_id`	The ID for the weather station where this data originated
`station_city`	The city where the weather station is located
`station_state`	A two-character code for the state where the weather station is located
`station_lat`	The latitude of this weather station
`station_lon`	The longitude of this weather station
`as_of_datetime`	The date and time that the data was gathered
`temp`	The temperature
`feels_like`	The temperature that it currently "feels like"
`wind`	The wind velocity (in kilometers per hour)
`wind_direction`	The direction of the wind
`precipitation`	Precipitation in the last hour (in millimeters)
`pressure`	Barometric pressure
`visibility`	The distance that can be clearly seen (in miles)
`humidity`	The percentage of relative humidity in the air
`weather_desc`	A text description of the current weather
`sunrise`	The time that the sun rises at this location today
`sunset`	The time that the sun sets at this location today

Approximately every hour, a CSV file containing the data for the locations that you requested will be delivered to you. The CSV file should look similar to Figure 16-2.

```
4589,Portland,ME,43.6591,70.2568,20240211 13:26,22,14,13,NNE,2.5,29.91,1.7,34,Heavy Snow,6:45,17:06
375,Boston,MA,42.3601,71.0589,20240211 13:27,24,15,11,NE,3.4,30.01,2.1,37,Snow,6:46,17:11
459,Providence,RI,41.8241,71.4128,20240211 13:26,25,15,11,SSW,3.1,27.32,1.7,38,Heavy Snow,6:47,17:14
778,New York,NY,40.7128,74.006,20240211 13:29,31,22,10,NE,2.2,29.83,3.3,34,Snow,6:55,17:26
4591,Philadelphia,PA,39.9526,75.1652,20240211 13:30,33,27,12,NW,2,29.85,5.7,88,Rain,6:58,17:32
753,Washington,DC,38.9072,77.0369,20240211 13:27,35,31,8,SSW,.3,30.51,8.1,74,Drizzle,7:04,17:41
507,Richmond,VA,37.5407,77.4361,20240211 13:28,43,38,10,S,0,28.14,9.1,64,Partly Cloudy,7:04,17:45
338,Raleigh,NC,35.7796,78.6382,20240211 13:27,52,51,4,ESE,0,29.33,9.2,56,Partly Sunny,7:06,17:52
759,Charleston,SC,32.7765,79.9311,20240211 13:28,61,59,6,W,0,29.74,9.5,54,Sunny,7:07,18:02
103,Jacksonville,FL,30.3322,81.6557,20240211 13:26,67,62,3,WSW,0,29.77,10,55,Sunny,7:10,18:12
2746,Miami,FL,25.7617,80.1918,20240211 13:28,76,78,1,SW,0,28.14,10,67,Sunny,6:59,18:12
```

Figure 16-2: The weather.csv *data file*

The file has one row for each of the 11 weather stations you requested, with every field delimited by a comma.

Creating the Weather Tables

Create a MySQL database called `weather` to store the weather data:

```
create database weather;
```

Now you'll create a table called `current_weather_load` to load the CSV file data into. The `_load` suffix makes it clear that this table is for loading data about the current weather.

Listing 16-1 shows the SQL statement to create the `current_weather_load` table.

```
create table current_weather_load
(
  station_id       int              primary key,
  station_city     varchar(100),
  station_state    char(2),
  station_lat      decimal(6,4)     not null,
  station_lon      decimal(7,4)     not null,
  as_of_dt         datetime,
  temp             int              not null,
  feels_like       int,
  wind             int,
  wind_direction   varchar(3),
  precipitation    decimal(3,1),
  pressure         decimal(6,2),
  visibility       decimal(3,1)     not null,
  humidity         int,
  weather_desc     varchar(100)     not null,
  sunrise          time,
  sunset           time,
  constraint check(station_lat between -90 and 90),
  constraint check(station_lon between -180 and 180),
  constraint check(as_of_dt between (now() - interval 1 day) and now()),
  constraint check(temp between -50 and 150),
```

```
  constraint check(feels_like between -50 and 150),
  constraint check(wind between 0 and 300),
  constraint check(station_lat between -90 and 90),
  constraint check(wind_direction in
    (
     'N','S','E','W','NE','NW','SE','SW',
     'NNE','ENE','ESE','SSE','SSW','WSW','WNW','NNW'
    )
  ),
  constraint check(precipitation between 0 and 400),
  constraint check(pressure between 0 and 1100),
  constraint check(visibility between 0 and 20),
  constraint check(humidity between 0 and 100)
);
```

Listing 16-1: Creating the current_weather_load table

Now create a second table called current_weather with the same structure as current_weather_load:

```
create table current_weather like current_weather_load;
```

With these two tables in place, you now have a table that you can load the CSV file to, as well as a final, user-facing table that you'll copy the weather data to once you are confident it has loaded cleanly.

Let's go over Listing 16-1 in more detail.

Data Types

You should always choose data types for the columns that match the data in the CSV file as closely as possible. For example, you define the station_id, temp, feels_like, wind, and humidity columns as int data types since they will come to you as numeric values without a decimal point. You define station_lat, station_lon, precipitation, pressure, and visibility as decimal data types because they will contain decimal points.

You should also consider how large the column values could be. For example, you define the station_lat column as decimal(6,4) because latitudes need to store numbers with up to two digits before the decimal point and four digits after the decimal point. You define station_lon as decimal(7,4) because it represents a longitude, which needs to store up to *three* digits before the decimal point and four digits after it. A longitude column needs to be able to hold a larger value than a latitude column.

You have to get creative with the as_of_dt column. Its data comes to you in the format YYYYMMDD hh:mm. MySQL doesn't have a data type that stores data in this format, so you create the as_of_dt column with a data type of datetime. When you load the data file into your load table, you'll convert this value to the datetime format. (We'll discuss how in the next section.)

The station_state column will always contain two characters, so you define it as char(2). Since the station_city and weather_desc columns will have a variable number of characters, you define both as a varchar containing up to 100 characters. No city or description should have more than

100 characters, so if you get a value for those columns that is larger than 100, you can safely say the data is incorrect.

The `sunrise` and `sunset` values come to you formatted as times with the hour and the minute provided. You use the `time` data type for those values, even though you aren't being sent the seconds in the data file. You'll load the values into columns with the `time` data type and let the seconds automatically default to zeros. For example, you'll load the value `17:06` and it will be saved in the table as `17:06:00`. This will work fine for your purposes since your application doesn't need to track the sunrise and sunset time down to the second.

Constraints

You create a primary key on the `station_id` column to enforce uniqueness. If the data file comes to you with two records for the same weather station, you don't want to load both records. Setting `station_id` as the primary key will prevent the second row from being loaded and will produce a warning message alerting you to a problem in the data file.

You add some other constraints to your columns as quality checks of the data that will be loaded into the table.

The `station_lat` column must be in the range of a valid latitude value: –90.0000 to 90.0000. You already defined `station_lat` with a data type of `decimal(6,4)` so there can be only six total digits, with four digits after the decimal point, but that won't prevent an invalid value like `95.5555` from being written to the column. Adding a `check` constraint will enforce that the value is in the appropriate range. This allows you to store all legitimate latitude values in your column and reject any values outside of that range. Similarly, the `station_lon` column must be in the range of a valid longitude value: –180.0000 to 180.0000.

The `wind_direction` column also has a `check` constraint to ensure that it contains only one of 16 possible values that you provided in a list (`N` for North, `SE` for Southeast, `NNW` for North-Northwest, and so on).

The other `check` constraints ensure that your data is within reasonable ranges for weather data. For example, a temperature outside of the range of –50 degrees to 150 degrees Fahrenheit is likely a mistake, so you'll reject it. Humidity is a percentage, so you enforce that it must be within the range of 0 to 100.

You also declare some of the columns in your load table with the `not null` constraint. These columns are so important that you want your load to fail if they are not provided. The `station_id` column must not be null since it is the primary key of the table.

You define `station_lat` and `station_lon` as `not null` because you want to plot the weather station's location on a map in your trucking application. You want to show each weather station's current temperature, visibility, and conditions at the right map location, and you can't do that if the station's latitude and longitude aren't provided.

The temperature, visibility, and weather_desc columns are also key pieces of data for this project, and thus you define them as not null as well.

Loading the Data File

Before you create the *weather.sh* Bash script that checks if a new CSV weather file is available, you'll write the *load_weather.sql* SQL script that will load the CSV file into your current_weather_load table (see Listing 16-2).

```
use weather;

delete from current_weather_load;

load data local infile '/home/weather_load/weather.csv'
into table current_weather_load
❶ fields terminated by ','
(
    station_id,
    station_city,
    station_state,
    station_lat,
    station_lon,
  ❷ @aod,
    temp,
    feels_like,
    wind,
    wind_direction,
    precipitation,
    pressure,
    visibility,
    humidity,
    weather_desc,
    sunrise,
    sunset
)
❸ set as_of_dt = str_to_date(@aod,'%Y%m%d %H:%i');

❹ show warnings;

❺ select concat('No data loaded for ',station_id,': ',station_city)
from    current_weather cw
where   cw.station_id not in
(
    select cwl.station_id
    from   current_weather_load cwl
);
```

Listing 16-2: The load_weather.sql *script*

First, you set your current database to the weather database and delete all rows from the current_weather_load table that may have been left over from a previous load.

Then you use the load data command you saw in Chapter 14 to load the *weather.csv* file into the current_weather_load table. Because you're loading a comma-separated file, you need to specify fields terminated by ',' ❶ so that load data knows where one field ends and the next field begins. You specify that the data file is called *weather.csv* and is in the */home/weather_load/* directory.

Within parentheses, you list all the columns in the table that you want the data fields loaded into, with one exception: instead of loading a value directly from the file into the as_of_dt column, you load it into a variable called @aod ❷. This variable holds the value of the as of date as it is formatted in the CSV file, which, as mentioned earlier, is YYYYMMDD hh:mm. You convert the value of the @aod variable from a string to a datetime data type using MySQL's str_to_date() function ❸. You use the format specifiers %Y, %m, %d, %H, and %i to specify the format of the string. By specifying str_to_date(@aod, '%Y%m%d %H:%i'), you're saying the @aod variable is made up of the following parts:

- %Y, a four-digit year
- %m, a two-digit month
- %d, a two-digit day
- A space
- %H, a two-digit hour (0–23)
- A colon
- %i, a two-digit minute (0–59)

With this information, the str_to_date() function has what it needs to convert the @aod string to the as_of_date datetime field in the current_weather_load table.

NOTE *The* str_to_date() *function can convert a string to a date,* time*, or* datetime *value, depending upon the format you provide. In this case, it returns a* datetime*.*

Next, you check if there were any problems loading the data. The show warnings command ❹ lists any errors, warnings, or notes from the last command that you ran. If problems in the data file caused your load data command to fail, show warnings will tell you what the problem was.

Then, you add a query as a second check that the data loaded properly ❺. In this query, you get a list of all the weather stations that had data the last time you wrote data into the current_weather table. If any of those weather stations aren't in the current_weather_load table you just loaded, that likely means weather station data was missing from your data file or there was a problem with that weather station's data that caused it not to load. In either case, you want to be notified.

You've now written your *load_weather.sql* script to notify you of any problems with loading the data. If *load_weather.sql* runs and no output is created, the data was loaded into the current_weather_load table without a problem.

Copying the Data to Your Final Table

Once the data is loaded from the CSV data file into your current_weather_load table without issue, you'll run another SQL script called *copy_weather.sql* to copy the data to your final current_weather table (Listing 16-3).

```
use weather;

delete from current_weather;

insert into current_weather
(
       station_id,
       station_city,
       station_state,
       station_lat,
       station_lon,
       as_of_dt,
       temp,
       feels_like,
       wind,
       wind_direction,
       precipitation,
       pressure,
       visibility,
       humidity,
       weather_desc,
       sunrise,
       sunset
)
select station_id,
       station_city,
       station_state,
       station_lat,
       station_lon,
       as_of_dt,
       temp,
       feels_like,
       wind,
       wind_direction,
       precipitation,
       pressure,
       visibility,
       humidity,
       weather_desc,
       sunrise,
       sunset
from   current_weather_load;
```

Listing 16-3: The copy_weather.sql *script*

This SQL script sets your current database to the weather database, deletes all old rows from the current_weather table, and loads the current_weather table with the data from the current_weather_load table.

Now that you have your SQL scripts written, you can write the Bash script that calls them (Listing 16-4).

```
#!/bin/bash ❶
cd /home/weather/ ❷

if [ ! -f weather.csv ]; then ❸
   exit 0
fi

mysql --local_infile=1 -h 127.0.0.1 -D weather -u trucking -pRoger -s \
       < load_weather.sql > load_weather.log ❹

if [ ! -s load_weather.log ]; then ❺
   mysql -h 127.0.0.1 -D weather -u trucking -pRoger -s < copy_weather.sql > copy_weather.log
fi

mv weather.csv weather.csv.$(date +%Y%m%d%H%M%S) ❻
```

Listing 16-4: The weather.sh *script*

The first line of a Bash script ❶ is known as a *shebang*. It tells the system that the interpreter to use for the commands in this file is in the */bin/bash* directory.

Next, you use the `cd` command to change directories to */home/weather/* ❷.

In your first `if` statement ❸, you check if the *weather.csv* file exists. In Bash scripts, `if` statements start with `if` and end with `fi`. The `-f` command checks if a file exists, and `!` is the syntax for `not`. The statement `if [ ! -f weather.csv ]` checks if the *weather.csv* file does not exist. If it doesn't, that means you don't have a new CSV data file to load, so you `exit` the Bash script and provide an exit code of `0`. By convention, you provide an exit code of `0` for success or `1` to signal an error. Exiting the Bash script here prevents the rest of the script from running; you don't need to run the rest of the script, since you don't have a data file to process.

You use the MySQL command line client ❹ (the `mysql` command, described in Chapter 14) to run the *load_weather.sql* SQL script. If the *load_weather.sql* script has any problems loading the data to the `current_weather_load` table, you'll log those problems to a file called *load_weather.log*.

In Bash, the left arrow (`<`) and right arrow (`>`) are used for *redirection*, which lets you take your input from a file and write your output to another file. The syntax `< load_weather.sql` tells the MySQL command line client to run the commands from the *load_weather.sql* script. The syntax `> load_weather.log` says to write any output to the *load_weather.log* file.

The `local_infile=1` option lets you run the `load data` command (used in the *load_weather.sql* script) using data files on your local computer, as opposed to data files on the server where MySQL is installed. This may be unnecessary in your environment, depending upon your configuration settings. (Your DBA can set this option as a configuration parameter using the command `set global local_infile = on`.)

The -h option tells the MySQL command line client which host server MySQL is installed on. In this case, -h 127.0.0.1 indicates that your MySQL host server is the same computer that you're currently using to run the script. Also known as *localhost*, 127.0.0.1 is the IP address for the current (local) computer. You could also simply type -h localhost here.

Next, you provide the database name, weather; your MySQL user ID, trucking; and your password, Roger. Oddly, MySQL doesn't allow a space after the -p, so enter your password without a preceding space.

You use the -s option to run your SQL script in *silent mode*. This prevents the script from giving you too much information in your output. For example, if no data gets loaded for the Boston weather station, you want to see the message No data loaded for 375: Boston in your *load_weather.log* file. But without the -s, the logfile will also show the beginning of the select statement that produced that message:

```
concat('No data loaded for ',station_id,': ',station_city)
No data loaded for 375: Boston
```

Using -s prevents the text concat('No data loaded for ',station_id,': ',station_city) from being written to *load_weather.log*.

In Bash, the backslash character (\) lets you continue your command on the next line. After the -s, you use a backslash and continue on the next line because your line of code was so long.

Next, your Bash script checks to see if any load problems are listed in the *load_weather.log* file ❺. In an if statement, -s checks to see if the file size is greater than 0 bytes. You only want to load the data into your final table, current_weather, if there were no problems loading the data into your load table, current_weather_load. In other words, you'll only copy the data to the current_weather table when the *load_weather.log* file is empty, or 0 bytes. You check that the logfile doesn't have a size greater than 0 using the syntax if [! -s load_weather.log].

Finally, in the last line of the *weather.sh* Bash script, you rename the *weather.csv* file, adding the current date and time as a suffix. For example, you'll rename *weather.csv* to *weather.csv.20240412210125* so that the next time your Bash script is run, it won't try to reload the same *weather.csv* file ❻. The mv command stands for *move*, and is used to rename or move files to another directory.

Now let's check out the results. If you are sent a *weather.csv* data file with valid data, running the *load_weather.sql* script will result in the current_weather_load table getting populated with values. This should look similar to Figure 16-3.

The data in your current_weather_load table looks good. All 11 rows that were in the CSV data file are now in the table, and the values look reasonable for all of your columns.

On the other hand, if you're sent a *weather.csv* data file with duplicate values, or with values that are in the wrong format or out of range, the result of running the *load_weather.sql* script will be that your *load_weather.log* file will contain a list of the problems.

station_id	station_city	station_state	station_lat	station_lon	as_of_dt	temp	feels_like	wind	wind_direction	precipitation	pressure	visibility	humidity	weather_desc	sunrise	sunset
4589	Portland	ME	43.6591	70.2568	2024-02-11 13:26:00	22	14	13	NNE	2.5	29.91	1.7	34	Heavy Snow	06:45:00	17:06:00
375	Boston	MA	42.3601	71.0589	2024-02-11 13:27:00	24	15	11	NE	3.4	30.01	2.1	37	Snow	06:46:00	17:11:00
459	Providence	RI	41.8241	71.4128	2024-02-11 13:26:00	25	15	11	SSW	3.1	27.32	1.7	38	Heavy Snow	06:47:00	17:14:00
778	New York	NY	40.7128	74.0060	2024-02-11 13:29:00	31	22	10	NE	2.2	29.83	3.3	34	Snow	06:55:00	17:26:00
4591	Philadelphia	PA	39.9526	75.1652	2024-02-11 13:30:00	33	27	12	NW	2.0	29.85	5.7	88	Rain	06:58:00	17:32:00
753	Washington	DC	38.9072	77.0369	2024-02-11 13:27:00	35	31	8	SSW	0.3	30.51	8.1	74	Drizzle	07:04:00	17:41:00
507	Richmond	VA	37.5407	77.4361	2024-02-11 13:28:00	43	38	10	S	0.0	28.14	9.1	64	Partly Cloudy	07:04:00	17:45:00
338	Raleigh	NC	35.7796	78.6382	2024-02-11 13:27:00	52	51	4	ESE	0.0	29.33	9.2	56	Partly Sunny	07:06:00	17:52:00
759	Charleston	SC	32.7765	79.9311	2024-02-11 13:28:00	61	59	6	W	0.0	29.74	9.5	54	Sunny	07:07:00	18:02:00
103	Jacksonville	FL	30.3322	81.6557	2024-02-11 13:26:00	67	62	3	WSW	0.0	29.77	10.0	55	Sunny	07:10:00	18:12:00
2746	Miami	FL	25.7617	80.1918	2024-02-11 13:28:00	76	78	1	SW	0.0	28.14	10.0	67	Sunny	06:59:00	18:12:00

Figure 16-3: The `current_weather_load` table

Assuming you got valid data and *copy_weather.sql* ran, the `current_weather` table should match Figure 16-3.

Next, you'll create the schedule to run this Bash script using cron.

Scheduling the Bash Script on cron

Using the command `crontab -e`, create the following crontab entry:

```
*/5 * * * * /home/weather/weather.sh
```

The `*/5` in the `minutes` column tells cron to run this job every 5 minutes. You can use the wildcard (*) character for all the other values (hour, day of month, month, and day of week, respectively), since you want the script to run for all hours, months, days, and days of the week. Figure 16-4 shows what each piece of the crontab entry means.

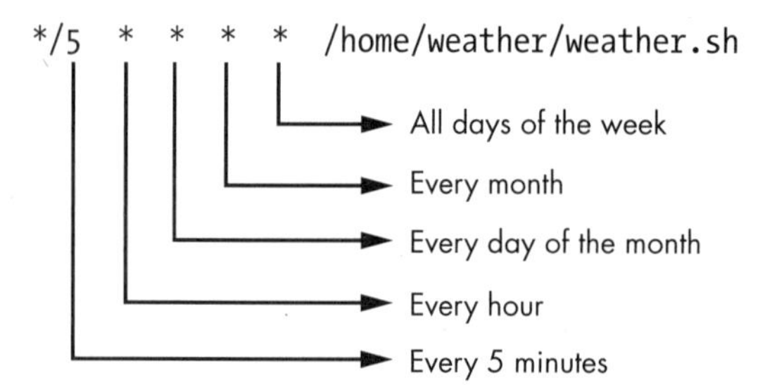

Figure 16-4: Scheduling weather.sh *on cron to run every 5 minutes*

You then save the crontab file and exit the text editor that was launched by the `crontab -e` command.

> **TRY IT YOURSELF**
>
> The *weather.csv* file contains valid weather data that will load into your `current_weather_load` table without a problem:
>
> ```
> 4589,Portland,ME,43.6591,70.2568,20240211 13:26,22,14,13,NNE,2.5,29.91,1.7
> ,34,Heavy Snow,6:45,17:06
> ```

```
375,Boston,MA,42.3601,71.0589,20240211 13:27,24,15,11,NE,3.4,30.01,2.1,37
,Snow,6:46,17:11
459,Providence,RI,41.8241,71.4128,20240211 13:26,25,15,11,SSW,3.1,27.32,1.
7,38,Heavy Snow,6:47,17:14
778,New York,NY,40.7128,74.006,20240211 13:29,31,22,10,NE,2.2,29.83,3.3,34
,Snow,6:55,17:26
4591,Philadelphia,PA,39.9526,75.1652,20240211 13:30,33,27,12,NW,2,29.85,5.
7,88,Rain,6:58,17:32
753,Washington,DC,38.9072,77.0369,20240211 13:27,35,31,8,SSW,.3,30.51,8.1,
74,Drizzle,7:04,17:41
507,Richmond,VA,37.5407,77.4361,20240211 13:28,43,38,10,S,0,28.14,9.1,64,P
artly Cloudy,7:04,17:45
338,Raleigh,NC,35.7796,78.6382,20240211 13:27,52,51,4,ESE,0,29.33,9.2,56,P
artly Sunny,7:06,17:52
759,Charleston,SC,32.7765,79.9311,20240211 13:28,61,59,6,W,0,29.74,9.5,54,
Sunny,7:07,18:02
103,Jacksonville,FL,30.3322,81.6557,20240211 13:26,67,62,3,WSW,0,29.77,10,
55,Sunny,7:10,18:12
2746,Miami,FL,25.7617,80.1918,20240211 13:28,76,78,1,SW,0,28.14,10,67,Su
nny,6:59,18:12
```

16-1. Edit the CSV file using a text editor and add some invalid data. Remove a line of data, duplicate a line of data, or replace a numeric value with a non-numeric value like an `X`. Rerun the *weather.sh* Bash script, either using cron or by changing directories to */home/weather/* and typing `./weather.sh` to manually run the script.

Check the contents of the *load_weather.log* file. Does it list the data issues that you put in the CSV file?

In the `weather` database, check the contents of the `current_weather_load` and `current_weather` tables using the `select * from` syntax. Does the `current_weather` table still contain the data from the last valid data load?

Alternative Approaches

As the saying goes, there are many ways to skin a cat. Likewise, there are many other ways you could have approached this project using what you've learned so far.

You could have loaded the data from the CSV file directly to the final `current_weather` table, but using an interim load table enables you to correct any data issues behind the scenes without affecting user-facing data. If the CSV file comes to you with data problems like duplicate records, incorrectly formatted column values, or out-of-range values, your load into the `current_weather_load` table will fail. While you work with the CSV file supplier to get a corrected file, your application will continue using the existing data in the `current_weather` table and your users won't be affected (though the weather data they see won't be as up to date as it normally would be).

If your weather data provider had an *application programming interface (API)* available, you could have received this weather data from an API rather than loading a CSV data file. An API is another way to exchange data between two systems, but an in-depth discussion is beyond the scope of this book.

You created a primary key and several other constraints on your `current_weather_load` table. You wouldn't do this in cases where you need to load a large number of records from a file into a table. For performance reasons, you'd choose to load the data into a table that has no constraints. As each row is being written into the table, MySQL needs to check that the constraints aren't being violated, which takes time. In your weather project, however, there were only 11 rows being loaded, so the load time was almost instantaneous even with the constraints.

You could have added a line of code to the Bash script, *weather.sh*, to have it notify you and the data provider by email or text whenever there's a problem loading the data. This wasn't included in the project because it requires a bit of setup. To learn more, use the `man` command to look up the `mailx`, `mail`, or `sendmail` commands (for example, `man mailx`).

Also, your database credentials are hardcoded in your *weather.sh* Bash script so that the script can call the MySQL command line client. When you load the data, MySQL gives you the warning `Using a password on the command line interface can be insecure`. It would be worth restructuring the code so that it hides your database user ID and password or using the `mysql_config_editor` utility shown in Chapter 14.

Summary

In this project, you scheduled a cron job to execute a Bash script that checks for the arrival of a CSV data file containing current weather data. When the file arrived, you loaded it into your MySQL database. You also checked for problems loading the data and, once it loaded cleanly, transferred the data to your final weather table.

In the next project, you'll use triggers to track changes to voter data in a MySQL database.

17

TRACKING CHANGES TO VOTER DATA WITH TRIGGERS

In this chapter, you'll build a voting database that stores data for an election. You'll improve the quality of your data by designing the database with constraints, including primary and foreign keys, and using triggers to prevent bad data from being entered. You'll also use triggers to track changes to your database so that if data quality issues arise, you have a record of who made the changes and when.

You'll allow poll workers to change data when appropriate, so it's important to build a system that prevents errors from being made. The techniques in this chapter can be applied to a wide variety of applications and situations. The quality of your data is crucial, so it's worth setting up your database in a way that keeps your data as accurate as possible.

NOTE *This is quite a large project, so you may want to approach it by tackling the sections on different days.*

Setting Up the Database

First, you'll create the database and take a look at its tables. The ballot for your election has races for mayor, treasurer, school committee, the board of health, and the planning board. Figure 17-1 shows the ballot you'll use for your database.

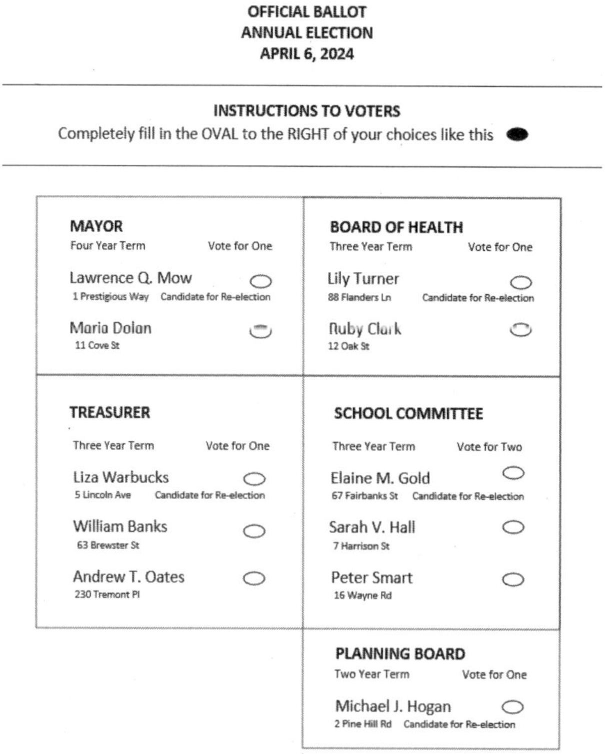

OFFICIAL BALLOT
ANNUAL ELECTION
APRIL 6, 2024

INSTRUCTIONS TO VOTERS
Completely fill in the OVAL to the RIGHT of your choices like this

MAYOR
Four Year Term Vote for One
Lawrence Q. Mow
1 Prestigious Way Candidate for Re-election
Maria Dolan
11 Cove St

BOARD OF HEALTH
Three Year Term Vote for One
Lily Turner
88 Flanders Ln Candidate for Re-election
Ruby Clark
12 Oak St

TREASURER
Three Year Term Vote for One
Liza Warbucks
5 Lincoln Ave Candidate for Re-election
William Banks
63 Brewster St
Andrew T. Oates
230 Tremont Pl

SCHOOL COMMITTEE
Three Year Term Vote for Two
Elaine M. Gold
67 Fairbanks St Candidate for Re-election
Sarah V. Hall
7 Harrison St
Peter Smart
16 Wayne Rd

PLANNING BOARD
Two Year Term Vote for One
Michael J. Hogan
2 Pine Hill Rd Candidate for Re-election

Figure 17-1: The ballot for your election

This election uses optical scan voting machines that read the ballots and save the voting data to your MySQL database.

Create the `voting` database:

```
create database voting;
```

Now you can begin adding tables.

Creating the Tables

You'll create the following tables within your `voting` database:

beer A table that contains data about beer.

voter People who are eligible to vote in this election

ballot The voter's ballot

race	The races on the ballot (for example, Mayor, Treasurer)
candidate	The candidates running
ballot_candidate	The candidates that the voter selected on their ballot

The *entity relationship diagram (ERD)* in Figure 17-2 shows these tables and their columns, as well as the primary and foreign key relationships between them.

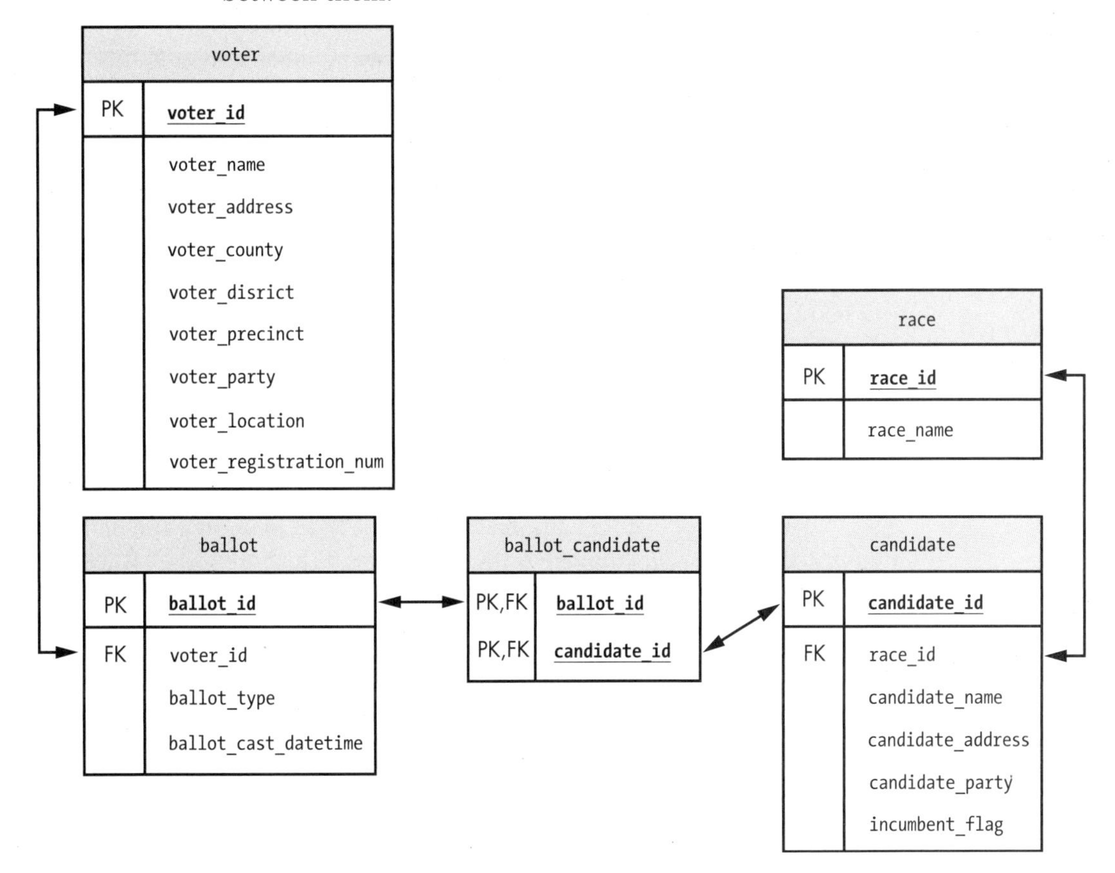

Figure 17-2: The tables in your voting database

Voters will cast a ballot with the candidates that they selected for each race.

The voter Table

The voter table will store information about each voter, such as name, address, and county. Create the table as follows:

```
use voting;

create table voter
    (
    voter_id                int             primary key     auto_increment,
```

```
    voter_name               varchar(100)     not null,
    voter_address            varchar(100)     not null,
    voter_county             varchar(50)      not null,
    voter_district           varchar(10)      not null,
    voter_precinct           varchar(10)      not null,
    voter_party              varchar(20),
    voting_location          varchar(100)     not null,
    voter_registration_num   int              not null        unique
    );
```

The voter_id column is the primary key for the table. Creating this primary key not only will speed up joins that use the voter table, but also will make sure that no two rows in the table have the same voter_id value.

You set voter_id to auto_increment so that MySQL will automatically increase the voter_id value with each new voter you add to the table.

You can't have two voters with the same registration number, so you set the voter_registration_num column to unique. If a new voter is added to the table with the same voter_registration_num as an existing voter, that new row will be rejected.

All of the columns are defined as not null except for the voter_party column. You'll allow a row to be saved in the table with a null voter_party, but if any other columns contain a null value, the row will be rejected.

The ballot Table

The ballot table holds information about each ballot, including the ballot number, the voter who completed the ballot, when the ballot was cast, and whether the ballot was cast in person or absentee. Create the ballot table like so:

```
create table ballot
    (
    ballot_id               int             primary key     auto_increment,
    voter_id                int             not null        unique,
    ballot_type             varchar(10)     not null,
    ballot_cast_datetime    datetime        not null        default now(),
    constraint foreign key (voter_id) references voter(voter_id),
    constraint check(ballot_type in ('in-person', 'absentee'))
    );
```

The ballot_id column is your primary key in this table, and its values automatically increment as you insert new ballot rows into the table.

You use a unique constraint for the voter_id column to ensure there is only one ballot in the table per voter. If a voter tries to cast more than one ballot, only the first ballot will be counted; subsequent ballots will be rejected.

The ballot_cast_datetime column saves the date and time that the ballot was cast. You set a default so that if a value isn't provided for this column, the now() function will write the current date and time to it.

You put a foreign key constraint on the ballot table's voter_id column to reject any ballots submitted by a voter whose information is not in the voter table.

Lastly, you add a check constraint to the ballot_type column that allows only the values in-person or absentee. Any rows with other ballot types will be rejected.

The race Table

The race table stores information about each race in your election, including the name of the race and how many candidates voters can vote for in it. You'll create it like so:

```
create table race
    (
    race_id               int             primary key     auto_increment,
    race_name             varchar(100)    not null        unique,
    votes_allowed         int             not null
    );
```

The race_id column is the primary key for this table, and is set to automatically increment. You define the race_name column with a unique constraint so that two races of the same name, like Treasurer, can't be inserted into the table.

The votes_allowed column holds the number of candidates voters can select in this race. For example, voters can choose one candidate for the mayoral race, and two for the school committee race.

The candidate Table

Next, you'll create the candidate table, which stores information about the candidates who are running:

```
create table candidate
    (
    candidate_id          int             primary key     auto_increment,
    race_id               int             not null,
    candidate_name        varchar(100)    not null        unique,
    candidate_address     varchar(100)    not null,
    candidate_party       varchar(20),
    incumbent_flag        bool,
    constraint foreign key (race_id) references race(race_id)
    );
```

The candidate_id column is the primary key for this table. This not only prevents duplicate rows from being entered by mistake, but also enforces a *business rule*—a requirement or policy about the way your system operates—that a candidate can run for only one race. For example, if a candidate tries to run for both mayor and treasurer, the second row would be rejected. You also define the candidate_id column to automatically increment.

The race_id column stores the ID of the race in which the candidate is running. The race_id is defined as a foreign key to the race_id column in the race table. This means that there can't be a race_id value in the candidate table that isn't also in the race table.

You define candidate_name as unique so that there can't be two candidates in the table with the same name.

> **THE ORDER OF TABLE CREATION**
>
> When you create tables that have foreign keys, the order of creation matters. For example, the candidate table has a foreign key column called race_id that references the race_id column of the race table. For that reason, you need to create the race table before you create the candidate table. If you try to create the candidate table first, you'll get an error:
>
> ```
> Error Code: 1824. Failed to open the referenced table 'race'
> ```
>
> MySQL is letting you know that you defined a foreign key that points to the race table, which doesn't exist yet.

The ballot_candidate Table

Now, you'll create your final table, ballot_candidate. This table tracks which candidates received votes on which ballot.

```
create table ballot_candidate
    (
    ballot_id       int,
    candidate_id    int,
    primary key (ballot_id, candidate_id),
    constraint foreign key (ballot_id) references ballot(ballot_id),
    constraint foreign key (candidate_id) references candidate(candidate_id)
    );
```

This is an associative table that references both the ballot and candidate tables. The primary key for this table comprises both the ballot_id and candidate_id columns. This enforces a rule that no candidate can get more than one vote from the same ballot. If someone attempted to insert a duplicate row with the same ballot_id and candidate_id, the row would be rejected. Both columns are also foreign keys. The ballot_id column is used to join to the ballot table and candidate_id is used to join to the candidate table.

By defining your tables with these constraints, you're improving the quality and integrity of the data in your database.

> **TRY IT YOURSELF**
>
> **17-1.** Create the voting database and its tables using the create database and create table statements shown previously.

17-2. Add a row to the voter table using this insert statement:

```
insert into voter
(
  voter_name,
  voter_address,
  voter_county,
  voter_district,
  voter_precinct,
  voter_party,
  voting_location,
  voter_registration_num
)
values
(
  'Susan King',
  '12 Pleasant St. Springfield',
  'Franklin',
  '12A',
  '4C',
  'Democrat',
  '523 Emerson St.',
  129756
);
```

17-3. Query the voter table using the `select * from voter;` syntax. Notice that the row you inserted in Exercise 17-2 has a value for the `voter_id` column. You didn't provide a `voter_id` value in your `insert` statement, so how do you think that value got there? Is there anything about the `create table` statement that would explain it?

17-4. Try to insert rows in the tables that violate your business rules. For example, insert a new voter row with the same `voter_registration_num` as an existing row in the voter table. Or insert a new row into the `ballot` table with a `voter_id` that doesn't exist in the voter table.

Are the constraints you defined when you created the tables working to prevent bad data from being entered?

Adding Triggers

You'll create several triggers on your tables to enforce business rules and track changes to your data for auditing purposes. These triggers will fire before or after a row is inserted, updated, or deleted from a table.

NOTE *In this chapter, you'll focus mostly on building the triggers for the* `voter` *and* `ballot` *tables. The code for the other triggers (relating to the* `race`, `candidate`, *and* `ballot_candidate` *tables) is similar and can be found on GitHub at* https://github.com/ricksilva/mysql_cc/blob/main/chapter_17.sql.

Before Triggers

You'll use triggers that fire *before* data gets changed to prevent data that doesn't adhere to your business rules from being written to your tables. In Chapter 12, you created a trigger that changed credit scores that were below 300 to exactly 300 right before the data was saved to a table. For this project, you'll use before triggers to make sure voters don't *overvote*, or vote for more candidates than allowed for that race. You'll also use before triggers to prevent particular users from making changes to some of your tables. Not every table will need a before trigger.

Business Rules

You'll enforce a few business rules using before triggers. First, although all poll workers are allowed to make changes to the `ballot` and `ballot_candidate` tables, only the secretary of state is allowed to make changes to data in the `voter`, `race`, and `candidate` tables. You'll create the following before triggers to enforce this business rule:

`tr_voter_bi`	Prevents other users from inserting voters
`tr_race_bi`	Prevents other users from inserting races
`tr_candidate_bi`	Prevents other users from inserting candidates
`tr_voter_bu`	Prevents other users from updating voters
`tr_race_bu`	Prevents other users from updating races
`tr_candidate_bu`	Prevents other users from updating candidates
`tr_voter_bd`	Prevents other users from deleting voters
`tr_race_bd`	Prevents other users from deleting races
`tr_candidate_bd`	Prevents other users from deleting candidates

These triggers will prevent users from making changes and will display an error message explaining that only the secretary of state is allowed to change this data.

Second, voters are allowed to select a certain number of candidates for each race. It's fine for voters to select no candidates for a race, or to select fewer than the maximum allowed number of candidates for a race, but they may not select more than the maximum number of candidates allowed. You'll prevent overvoting by creating the `tr_ballot_candidate_bi` trigger.

These are all the before triggers you'll need for this project. Remember, some tables won't have before triggers.

Before Insert Triggers

You'll need four *before insert* triggers for your project. Three of them will prevent users other than the secretary of state from inserting data in your `voter`, `race`, and `candidate` tables. The other before insert trigger will prevent voters from voting for too many candidates in a race.

In Listing 17-1, you write the before insert trigger to prevent users other than the secretary of state from inserting new rows in your `voter` table.

```
drop trigger if exists tr_voter_bi;

delimiter //

create trigger tr_voter_bi
  before insert on voter
  for each row
begin
  if user() not like 'secretary_of_state%' then
❶ signal sqlstate '45000'
    set message_text = 'Voters can be added only by the Secretary of State';
  end if;
end//

delimiter ;
```

Listing 17-1: Defining the tr_voter_bi trigger

First, in case the trigger already exists, you drop it before you re-create it. You define the `tr_voter_bi` trigger as a `before insert` trigger. For each row being inserted into the `voter` table, you check that the name of the user inserting the new voter starts with the text `secretary_of_state`.

The `user()` function returns both the username and the hostname, like `secretary_of_state@localhost`. If that string doesn't start with the text `secretary_of_state`, it means somebody other than the secretary of state is trying to insert a voter record. In that case, you'll send an error message with the `signal` statement ❶.

You might remember from Chapter 12 that `sqlstate` is a five-character code that identifies errors and warnings. The value you used, `45000`, is an error condition that causes your trigger to exit. This prevents the row from being written to the `voter` table.

You can define the message to display by using the `set message_text` syntax. Notice that this line is a part of the `signal` command, as there is no semicolon at the end of the `signal` line. You could have combined these two lines into one, like this:

```
signal sqlstate '45000' set message_text = 'Voters can be added only...';
```

This `tr_voter_bi` trigger prevents users other than the secretary of state from inserting voter rows.

TRY IT YOURSELF

17-5. Write similar before insert triggers for the `race` and `candidate` tables to prevent users other than the secretary of state from inserting new races and candidates.

(continued)

To test these triggers, first log in to MySQL Workbench and insert a row into the tables. For example, for the `race` table, you could run this SQL:

```
insert into race (race_name, votes_allowed)
values ('Dog Catcher', 1);
```

You should get the error message `Voters can be added only by the Secretary of State.`

Create a new MySQL user for the secretary of state like so:

```
create user secretary_of_state@localhost identified by 'v0t3';
```

This creates the `secretary_of_state@localhost` user with a password of `v0t3`.

Grant the new user permissions using this command:

```
grant all privileges on *.* to secretary_of_state@localhost;
```

(Granting superuser privileges on everything is normally a bad idea, but we'll do it just for this test.)

Now create a MySQL Workbench connection using `secretary_of_state@localhost` as shown in the following figure.

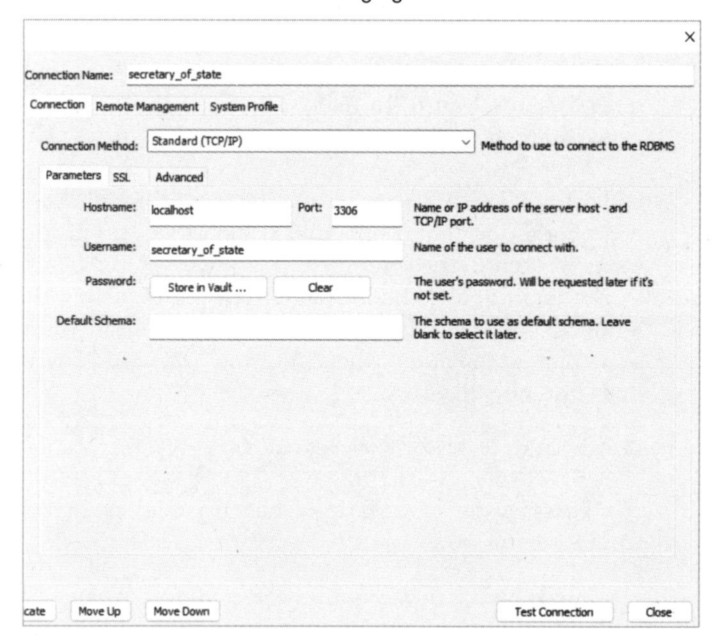

Run the SQL to add a new `race`:

```
insert into race (race_name, votes_allowed)
values ('Dog Catcher', 1);
```

Now, because you're running the `insert` statement using the secretary of state's username, inserting the new race should succeed.

Now, write your tr_ballot_candidate_bi trigger to prevent voters from voting for too many candidates in a race (Listing 17-2).

```
drop trigger if exists tr_ballot_candidate_bi;

delimiter //

create trigger tr_ballot_candidate_bi
  before insert on ballot_candidate
  for each row
begin
  declare v_race_id int;
  declare v_votes_allowed int;
  declare v_existing_votes int;
  declare v_error_msg varchar(100);
  declare v_race_name varchar(100);

❶ select r.race_id,
         r.race_name,
         r.votes_allowed
❷ into   v_race_id,
         v_race_name,
         v_votes_allowed
  from   race r
  join   candidate c
  on     r.race_id = c.race_id
  where  c.candidate_id = new.candidate_id;

❸ select count(*)
  into   v_existing_votes
  from   ballot_candidate bc
  join   candidate c
  on     bc.candidate_id = c.candidate_id
  and    c.race_id = v_race_id
  where  bc.ballot_id = new.ballot_id;

  if v_existing_votes >= v_votes_allowed then
     select concat('Overvoting error: The ',
              v_race_name,
              ' race allows selecting a maximum of ',
              v_votes_allowed,
              ' candidate(s) per ballot.'
            )
     into v_error_msg;

  ❹ signal sqlstate '45000' set message_text = v_error_msg;
  end if;
end//

delimiter ;
```

Listing 17-2: Defining the `tr_ballot_candidate_bi` trigger

Before a new row is inserted into the ballot_candidate table, your trigger finds the number of votes allowed for that race. Then, it checks how many existing rows are in the ballot_candidate table for this ballot and this race. If the number of existing votes is greater than or equal to the maximum allowed, the new row is prevented from being inserted. (The number of existing votes should never be greater than the maximum allowed, but you'll check just for completeness.)

You declare five variables in your trigger: v_race_id holds the race ID, v_race_name holds the name of the race, v_existing_votes stores the number of votes that have already been cast on this ballot for candidates in this race, v_votes_allowed holds the number of candidates that voters are allowed to select in this race, and the v_error_msg variable holds an error message to display to the user in case too many candidates have been selected.

NOTE *In this example, you've declared each variable on its own line, but you could also declare groups of variables that have the same data types like so:*

```
declare v_race_id, v_votes_allowed, v_existing_votes int;
declare v_error_msg, v_race_name varchar(100);
```

Notice that all of the int *variables are declared together on the same line, and all of the* varchar(100) *variables are declared together.*

In the first select statement ❶, you use the candidate_id that is about to be inserted in the table—new.candidate_id—to get information about the race the candidate is running for. You join to the race table and get the race_id, race_name, and votes_allowed for the race and save them to variables ❷.

In your second select statement, you get a count of how many votes already exist in the ballot_candidate table for this race and this ballot ❸. You join to the candidate table to get the list of candidates that are running for this race. Then you count the number of rows in the ballot_candidate table with a row that has one of those candidates and this ballot ID.

If the ballot_candidate table already has the maximum number of votes for this ballot and this race, you'll use the signal command with a sqlstate code of 45000 to exit from the trigger and prevent the new row from being written to the ballot_candidate table ❹. You'll display the error message that you stored in the v_error_msg variable to the user:

```
Overvoting error: The Mayor race allows selecting a maximum of 1 candidate(s)
per ballot.
```

Before Update Triggers

You also need to prevent users other than the secretary of state from updating voter rows by writing a tr_voter_bu trigger, as shown in Listing 17-3.

```
drop trigger if exists tr_voter_bu;

delimiter //

create trigger tr_voter_bu
  before update on voter
```

```
  for each row
begin
  if user() not like 'secretary_of_state%' then
    signal sqlstate '45000'
    set message_text = 'Voters can be updated only by the Secretary of State';
  end if;
end//

delimiter ;
```

Listing 17-3: Defining the `tr_voter_bu` trigger

This trigger will fire before a row is updated in the `voter` table.

Although the before insert and before update triggers are similar, there is no way to combine them into one trigger. MySQL doesn't have a way to write a `before insert or update` trigger; it requires you to write two separate triggers instead. You can, however, call stored procedures from triggers. If two triggers shared similar functionality, you could add that functionality to a stored procedure and have each trigger call that procedure.

> **TRY IT YOURSELF**
>
> **17-6.** Using the `tr_voter_bu` trigger as a model, write the before update triggers for the `race` and `candidate` tables.

Before Delete Triggers

Next you'll write the `tr_voter_bd` trigger to prevent any user other than the secretary of state from deleting voter data (Listing 17-4).

```
drop trigger if exists tr_voter_bd;

delimiter //

create trigger tr_voter_bd
  before delete on voter
  for each row
begin
  if user() not like 'secretary_of_state%' then
    signal sqlstate '45000'
    set message_text = 'Voters can be deleted only by the Secretary of State';
  end if;
end//

delimiter ;
```

Listing 17-4: Defining the `tr_voter_bd` trigger

> **TRY IT YOURSELF**
>
> **17-7.** Using the `tr_voter_bd` trigger as a model, write the before delete triggers for the `race` and `candidate` tables.

After Triggers

You'll be writing triggers that fire *after* your data is inserted, updated, or deleted to track the changes made to your tables. But since the purpose of after triggers is to write rows to the audit tables, you need to create those audit tables first. These audit tables save a record of the changes made to the data in your tables, similar to those you saw in Chapter 12.

Audit Tables

Name your audit tables with the `_audit` suffix. For example, you'll track changes made to the `voter` table in the `voter_audit` table. You'll name all audit tables this way so it's clear what data they're tracking.

Create the audit tables as shown in Listing 17-5.

```
create table voter_audit
(
  audit_datetime  datetime,
  audit_user      varchar(100),
  audit_change    varchar(1000)
);

create table ballot_audit
(
  audit_datetime  datetime,
  audit_user      varchar(100),
  audit_change    varchar(1000)
);

create table race_audit
(
  audit_datetime  datetime,
  audit_user      varchar(100),
  audit_change    varchar(1000)
);

create table candidate_audit
(
  audit_datetime  datetime,
  audit_user      varchar(100),
  audit_change    varchar(1000)
);
```

```
create table ballot_candidate_audit
(
  audit_datetime  datetime,
  audit_user      varchar(100),
  audit_change    varchar(1000)
);
```

Listing 17-5: Creating audit tables before defining your after triggers

All of your audit tables are defined with the same structure. Each table has an audit_datetime column that contains the date and time that the change was made, an audit_user column that contains the name of the user who made the changes, and an audit_change column that contains a description of the data that was changed. When you find data in your voting application that doesn't seem right, you can look to these audit tables to find out more information.

Next, for each data table you'll create three after triggers that fire after an insert, update, or delete. The names of the triggers are shown in Table 17-1.

Table 17-1: After Trigger Names

Table	After insert triggers	After update triggers	After delete triggers
voter	tr_voter_ai	tr_voter_au	tr_voter_ad
ballot	tr_ballot_ai	tr_ballot_au	tr_ballot_ad
race	tr_race_ai	tr_race_au	tr_race_ad
candidate	tr_candidate_ai	tr_candidate_au	tr_candidate_ad
ballot_candidate	tr_ballot_candidate_ai	tr_ballot_candidate_au	tr_ballot_candidate_ad

You'll start with the after_insert trigger for each table.

After Insert Triggers

The tr_voter_ai trigger will fire after new rows are inserted into the voter table, adding rows to the voter_audit table to track the new data (see Listing 17-6).

```
drop trigger if exists tr_voter_ai;

delimiter //

create trigger tr_voter_ai
  after insert on voter
❶ for each row
begin
  insert into voter_audit
  (
    audit_datetime,
    audit_user,
    audit_change
```

```
)
values
(
❷ now(),
  user(),
  concat(
  ❸ 'New Voter added -',
      ' voter_id: ',                    new.voter_id,
      ' voter_name: ',                  new.voter_name,
      ' voter_address: ',               new.voter_address,
      ' voter_county: ',                new.voter_county,
      ' voter_district: ',              new.voter_district,
      ' voter_precinct: ',              new.voter_precinct,
      ' voter_party: ',                 new.voter_party,
      ' voting_location: ',             new.voting_location,
      ' voter_registration_num: ',      new.voter_registration_num
  )
 );
end//

delimiter ;
```

Listing 17-6: Defining the `tr_voter_ai` *trigger*

To create the trigger, you first check if the `tr_voter_ai` trigger already exists. If so, you drop it before re-creating it. Since a SQL `insert` statement can insert one row or many rows, you specify that for each row being inserted into the `voter` table, you want to write a single row to the `voter_audit` table ❶.

In the `audit_datetime` column, you insert the current date and time using the `now()` function ❷. In the `audit_user` column, you use the `user()` function to insert the name of the user who made the change. The `user()` function also returns the user's hostname, so usernames are followed by an at sign (`@`) and a hostname, like `clerk_238@localhost`.

You use the `concat()` function in the `audit_change` column to build a string that shows the values that were inserted. You start with the text `New voter added -` ❸ and get the inserted values by using the `new` keyword that's available to you in `insert` triggers. For example, `new.voter_id` shows you the `voter_id` that was just inserted into the `voter` table.

After a new row is added to the `voter` table, the `tr_voter_ai` trigger fires and writes a row with values like the following to the `voter_audit` table:

```
audit_datetime:  2024-05-04 14:13:04

audit_user:      secretary_of_state@localhost

audit_change:    New voter added - voter_id: 1 voter_name: Susan King
                 voter_address: 12 Pleasant St. Springfield
                 voter_county: Franklin voter_district: 12A voter_precinct: 4C
                 voter_party: Democrat voting_location: 523 Emerson St.
                 voter_registration_num: 129756
```

The trigger writes the datetime, user (and hostname), and details about the new voter to the audit table.

TRY IT YOURSELF

17-8. Using the tr_voter_ai trigger as a model, create the after insert trigger for the ballot table. The trigger should be called tr_ballot_ai and write to the ballot_audit table.

The trigger should use the new keyword to get the values for the columns. The columns in the ballot table are ballot_id, voter_id, ballot_type, and ballot_cast_datetime.

After you create the trigger, test it by inserting a new row into the ballot table, like so:

```
insert into ballot
(
    voter_id,
    ballot_type,
    ballot_cast_datetime
)
values
(
    1,
    'in-person',
    now()
);
```

For this insert statement to work, there must first be a row with a voter_id of 1 in the voter table, because voter_id in the ballot table is a foreign key that references voter_id in the voter table. For this reason, you need to do Exercise 17-2 before you do this exercise to insert that voter row.

Did the new row you inserted into the ballot table get logged to the ballot_audit table? Query the ballot_audit table by typing **select * from ballot_audit;** and see if you get the results that you expected.

17-9. Write the after insert triggers for the race, candidate, and ballot_candidate tables. These triggers will be similar to the ones you've already written. You just need to change the trigger name, the data table name, the audit table name, and the list of columns.

To compare your code to the completed code in GitHub, go to *https://github.com/ricksilva/mysql_cc/blob/main/chapter_17.sql.*

After Delete Triggers

In Listing 17-7 you write the after delete trigger, called tr_voter_ad, which will fire after rows are deleted from the voter table and track the deletions in the voter_audit table.

```
drop trigger if exists tr_voter_ad;

delimiter //
```

```
create trigger tr_voter_ad
❶ after delete on voter
  for each row
begin
  insert into voter_audit
  (
    audit_datetime,
    audit_user,
    audit_change
  )
  values
  (
    now(),
    user(),
    concat(
      'Voter deleted -',
         ' voter_id: ',                    old.voter_id,
         ' voter_name: ',                  old.voter_name,
         ' voter_address: ',               old.voter_address,
         ' voter_county: ',                old.voter_county,
         ' voter_district: ',              old.voter_district,
         ' voter_precinct: ',              old.voter_precinct,
         ' voter_party: ',                 old.voter_party,
         ' voting_location: ',             old.voting_location,
         ' voter_registration_num: ',      old.voter_registration_num
    )
  );
end//

delimiter ;
```

Listing 17-7: Defining the `tr_voter_ad` *trigger*

You define this trigger as `after delete` on the `voter` table ❶. You use the `user()` and `now()` functions to get the user who deleted the `voter` row and the date and time at which the row was deleted. You build a string, using the `concat()` function, that shows the values that were deleted.

The after delete trigger looks similar to your after insert trigger, but you use the `old` keyword instead of `new`. You can precede your column names with `old` and a period to get their value. For example, use `old.voter_id` to get the value of the `voter_id` column for the row that was just deleted.

After a row is deleted from the `voter` table, the `tr_voter_ad` trigger fires and writes a row to the `voter_audit` table with values like the following:

```
audit_datetime:  2024-05-04 14:28:54

audit_user:      secretary_of_state@localhost

audit_change:    Voter deleted - voter_id: 87 voter_name: Ruth Bain
                 voter_address: 887 Wyoming St. Centerville
                 voter_county: Franklin voter_district: 12A voter_precinct: 4C
                 voter_party: Republican voting_location: 523 Emerson St.
                 voter_registration_num: 45796
```

The trigger writes the datetime, user (and hostname), and details about the deleted voter record to the audit table.

> **TRY IT YOURSELF**
>
> **17-10.** Using the tr_voter_ad trigger as a model, create the after delete trigger for the ballot table. The trigger should be called tr_ballot_ad and should write to the ballot_audit table.
>
> After you create the trigger, test it by deleting a row from the ballot table, like this:
>
> ```
> delete from ballot
> where ballot_id = 1;
> ```
>
> Was the row that was deleted from the ballot table logged to the ballot_audit table? When you run the select * from ballot_audit query, do you see a record of the deleted row?
>
> **17-11.** Create the after delete triggers for the other tables in your voter database. The triggers should be named tr_race_ad, tr_candidate_ad, and tr_ballot_candidate_ad. The triggers should write to the race_audit, candidate_audit, and ballot_candidate_audit tables, respectively.
>
> You can test the triggers by deleting rows from the race, candidate, and ballot_candidate tables. When you query the audit tables, do you see a record of the deleted rows?

After Update Triggers

Now you'll write the after update trigger, tr_voter_au, which will fire after rows in the voter table are updated and track the change in the voter_audit table (Listing 17-8).

```
drop trigger if exists tr_voter_au;

delimiter //

create trigger tr_voter_au
  after update on voter
  for each row
begin
  set @change_msg = concat('Voter ',old.voter_id,' updated: ');

❶ if (new.voter_name != old.voter_name) then
  ❷ set @change_msg =
        concat(
            @change_msg,
            'Voter name changed from ',
            old.voter_name,
            ' to ',
```

```
            new.voter_name
        );
end if;

if (new.voter_address != old.voter_address) then
  set @change_msg =
      concat(
          @change_msg,
          '. Voter address changed from ',
          old.voter_address,
          ' to ',
          new.voter_address
  );
end if;

if (new.voter_county != old.voter_county) then
  set @change_msg =
      concat(
          @change_msg,
          '. Voter county changed from ', old.voter_county, ' to ',
          new.voter_county
      );
end if;

if (new.voter_district != old.voter_district) then
  set @change_msg =
      concat(
          @change_msg,
          '. Voter district changed from ',
          old.voter_district,
          ' to ',
          new.voter_district
      );
end if;

if (new.voter_precinct != old.voter_precinct) then
  set @change_msg =
      concat(
          @change_msg,
          '. Voter precinct changed from ',
          old.voter_precinct,
          ' to ',
          new.voter_precinct
      );
end if;

if (new.voter_party != old.voter_party) then
  set @change_msg =
      concat(
          @change_msg,
          '. Voter party changed from ',
          old.voter_party,
          ' to ',
```

```
            new.voter_party
        );
  end if;

  if (new.voting_location != old.voting_location) then
    set @change_msg =
        concat(
            @change_msg,
            '. Voting location changed from ',
            old.voting_location, '
            to ',
            new.voting_location
        );
  end if;

  if (new.voter_registration_num != old.voter_registration_num) then
     set @change_msg =
         concat(
             @change_msg,
             '. Voter registration number changed from ',
             old.voter_registration_num,
             ' to ',
             new.voter_registration_num
         );
  end if;

insert into voter_audit(
    audit_datetime,
    audit_user,
    audit_change
    )
values (
    now(),
    user(),
  ❸ @change_msg
    );

end//

delimiter ;
```

Listing 17-8: Defining the tr_voter_au trigger

Because the after update trigger fires after a row gets updated in a table, it can take advantage of both the new and old keywords. For example, you can see if the voter_name column value was updated in the voter table by checking new.voter_name != old.voter_name ❶. If the new value of the voter's name isn't the same as the old value, it was updated, and you'll save that information to write to the audit table.

For your insert and delete triggers, you wrote the values for *all* the columns in the voter table to the voter_audit table, but for your update trigger, you'll report only on the column values that changed.

For example, if you ran this update statement

```
update voter
set    voter_name = 'Leah Banks-Kennedy',
       voter_party = 'Democrat'
where  voter_id = 5876;
```

your update trigger would write a row to the voter_audit table with just these changes:

```
audit_datetime:  2024-05-08 11:08:04

audit_user:      secretary_of_state@localhost

audit_change:    Voter 5876 updated: Voter name changed from Leah Banks
                 to Leah Banks-Kennedy. Voter party changed from Republican
                 to Democrat
```

Since there were only two column values that changed, voter_name and voter_party, you'll write those two changes to your audit table.

To capture the changes that were made, you create a variable called @change_msg ❷. Using if statements, you check if each column value changed. When a column's value has changed, you use the concat() function to add information about that column's changes to the end of the existing @change_msg string variable. Once you've checked all of the column values for changes, you write the value of @change_msg variable to the audit_change column of the audit table ❸. You also write to the audit table the username of the person who made the change to the audit_user column, and the date and time that the change was made to the audit_datetime column.

TRY IT YOURSELF

17-12. Using the tr_voter_au trigger as a model, create the after update trigger for the ballot table. The trigger should be called tr_ballot_au and should write to the ballot_audit table.

After you create the trigger, you can test it by updating a row in the ballot table:

```
update ballot
set    ballot_type = 'absentee'
where  ballot_id = 1;
```

Was the value that was updated in the ballot table logged to the ballot_audit table? When you run the select * from ballot_audit query, do you see a record of the updated value?

17-13. Create the after update triggers for the other tables in your voter database. The triggers should be named `tr_race_au`, `tr_candidate_au`, and `tr_ballot_candidate_au`. The triggers should write to the `race_audit`, `candidate_audit`, and `ballot_candidate_audit` tables, respectively.

You've successfully built a database that not only stores your election data, but also includes constraints and triggers that keep the data at a high quality.

Alternative Approaches

As with the `weather` database project in the previous chapter, there are numerous approaches to writing this `voter` database.

Audit Tables

In this project, you created five different audit tables. Instead, you could have created just one audit table and written all of the audit records there. Alternatively, you could have created 15 audit tables: three for each table. For example, rather than auditing voter inserts, deletes, and updates to the `voter_audit` table, you could have audited new voters to a table called `voter_audit_insert`, changes to voters to `voter_audit_update`, and deletions to `voter_audit_delete`.

Triggers vs. Privileges

Rather than using triggers to control which users can update which tables, your database administrator could have done this by granting and revoking these privileges to and from your database users. The advantage of using triggers is that you're able to display a customized message to the user explaining the problem, like `Voters can be added only by the Secretary of State`.

Replacing check Constraints with New Tables

When you created the `ballot` table, you used the following `check` constraint to make sure that the `ballot_type` column has a value of `in-person` or `absentee`:

```
constraint check(ballot_type in ('in-person', 'absentee'))
```

Another approach would have been to create a `ballot_type` table that has rows for each ballot type, like this:

```
ballot_type_id  ballot_type
--------------  ---------------
       1        in-person
       2        absentee
```

You could have added a table, named ballot_type, and made the ballot_type_id column the primary key. If you did, you would save the ballot_type_id instead of the ballot_type in the ballot table. This would look like Figure 17-3.

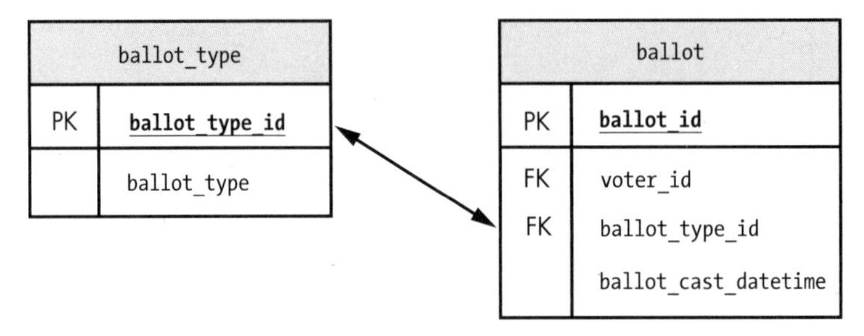

Figure 17-3: Creating a ballot_type table to store ballot types

One advantage to this approach is that you could add new ballot types, like military or overseas, without having to change the definition of the ballot table. It's also more efficient for each row of the ballot table to save an ID representing the ballot type, like 3, rather than saving the full name, like absentee.

You could have used a similar approach for the voter table. Instead of creating the voter table with the columns voter_county, voter_district, voter_precinct, and voter_party, you could have built the table to save just the IDs: voter_county_id, voter_district_id, voter_precinct_id, and voter_party_id and referenced new tables named county, district, precinct, and party to get the list of valid IDs.

There is plenty of room for creativity when creating databases, so don't feel as though you need to strictly follow the approach I've used in this project. Try any of these alternative approaches and see how they work for you!

Summary

In this chapter, you built a voting database that stores data for an election. You prevented data integrity problems using constraints and triggers, and tracked changes to your data using audit tables. You also saw some possible alternative approaches to this project. In the third and final project, you'll use views to hide sensitive salary data.

18

PROTECTING SALARY DATA WITH VIEWS

In this project, you'll use views to hide sensitive salary data in an employee table. The company in question has one database user from each department (Human Resources, Marketing, Accounting, Technology, and Legal) who is allowed access to most employee data. However, only users from Human Resources should be able to access the employees' salaries.

Views can hide sensitive data, but they can also be used to simplify access to a complex query, or to select just the relevant data in a table—for example, to show just the table's rows for a particular department.

Creating the employee Table

Start by creating your business database:

```
create database business;
```

Next, create an employee table that stores information about each employee in the company, including full name, job title, and salary:

```
use business;

create table employee
(
  employee_id       int               primary key    auto_increment,
  first_name        varchar(100)      not null,
  last_name         varchar(100)      not null,
  department        varchar(100)      not null,
  job_title         varchar(100)      not null,
  salary            decimal(15,2)     not null
);
```

Since you created the employee_id column as auto_increment, you don't need to provide an employee_id value when inserting new rows into the employee table. MySQL keeps track of that for you, and makes sure that the employee_id value gets higher with each row you insert. Add the following data to your table:

```
insert into employee(first_name, last_name, department, job_title, salary)
values ('Jean',' Unger', 'Accounting', 'Bookkeeper', 81200);

insert into employee(first_name, last_name, department, job_title, salary)
values ('Brock', 'Warren', 'Accounting', 'CFO', 246000);

insert into employee(first_name, last_name, department, job_title, salary)
values ('Ruth', 'Zito', 'Marketing', 'Creative Director', 178000);

insert into employee(first_name, last_name, department, job_title, salary)
values ('Ann', 'Ellis', 'Technology', 'Programmer', 119500);

insert into employee(first_name, last_name, department, job_title, salary)
values ('Todd', 'Lynch', 'Legal', 'Compliance Manager', 157000);
```

Now, query the table to see the inserted rows:

```
select * from employee;
```

The result is as follows:

```
employee_id  first_name  last_name  department  job_title           salary
-----------  ----------  ---------  ----------  ------------------  ---------
1            Jean        Unger      Accounting  Bookkeeper           81200.00
2            Brock       Warren     Accounting  CFO                 246000.00
3            Ruth        Zito       Marketing   Creative Director   178000.00
4            Ann         Ellis      Technology  Programmer          119500.00
5            Todd        Lynch      Legal       Compliance Manager  157000.00
```

The employee table data looks good, but you want to hide the salary column from everyone except the Human Resources user so that coworkers can't access one another's sensitive information.

Creating the View

Instead of allowing all database users to access the employee table, you'll let them access a view called v_employee that has the columns from the employee table minus the salary column. As discussed in Chapter 10, a view is a virtual table based on a query. Create the view like so:

```
create view v_employee as
select employee_id,
       first_name,
       last_name,
       department,
       job_title
from   employee;
```

You've left out the salary column from the select statement, so it shouldn't appear in your result once you query your view:

```
select * from v_employee;
```

The result is as follows:

```
employee_id first_name last_name department      job_title
----------- ---------- --------- ----------  ------------------
1           Jean       Unger     Accounting  Bookkeeper
2           Brock      Warren    Accounting  CFO
3           Ruth       Zito      Marketing   Creative Director
4           Ann        Ellis     Technology  Programmer
5           Todd       Lynch     Legal       Compliance Manager
```

As expected, the v_employee view contains every column except for salary.

Next, you'll change the permissions of the employee database to allow Human Resources to make changes in the underlying employee table. Since v_employee is a view, the changes to employee will be immediately reflected there.

Controlling Permissions

To adjust the permissions in your database, you'll use the grant command, which grants privileges to MySQL database users and controls which users can access which tables.

You have one database user per department: accounting_user, marketing_user, legal_user, technology_user, and hr_user. Grant access to the employee table to only hr_user by entering the following:

```
grant select, delete, insert, update on business.employee to hr_user;
```

You've granted hr_user the ability to select, delete, insert, and update rows in the employee table in the business database. You won't grant that access to the users from other departments. For example, if accounting_user tries to query the employee table, they'll get the following error message:

```
Error Code: 1142. SELECT command denied to user 'accounting_user'@'localhost'
for table 'employee'
```

Now you'll grant select access to your v_employee view to your users from all of your departments:

```
grant select on business.v_employee to hr_user@localhost;
grant select on business.v_employee to accounting_user@localhost;
grant select on business.v_employee to marketing_user@localhost;
grant select on business.v_employee to legal_user@localhost;
grant select on business.v_employee to technology_user@localhost;
```

All of your departments' users can select from the v_employee view to access the employee data they need.

REVOKING PRIVILEGES

If a user has already been granted access that you want to take away, you (or your DBA) can use the revoke command, like so:

```
revoke select, delete, insert, update
on business.employee from legal_user@localhost;
```

This command revokes select, delete, insert, and update access on the employee table for legal_user.

For this project, you can grant privileges using the root superuser account that was created when you installed MySQL (see Chapter 1). In a live production environment, your DBA would typically create other accounts rather than using root, which has all privileges and can do anything. In a professional setting, very few people know the root password. A DBA can also define permissions to a *role* and then add or remove users as members of that role, but a detailed discussion of roles is beyond the scope of this book.

Using MySQL Workbench to Test User Access

You'll use MySQL Workbench with this project and connect as root to create the database, tables, and departments' users. Then, you'll create separate connections as hr_user and accounting_user to see how their access differs.

NOTE *You could use another tool, like the MySQL command line client or MySQL Shell, but I like the ease with which you can test different users' access by clicking a MySQL Workbench connection and logging in as that user.*

First, create a connection for the `root` user, using the password that you created when you installed MySQL. To create the connection, click the + icon next to the text MySQL Connections on the Welcome to MySQL Workbench screen, as shown in Figure 18-1.

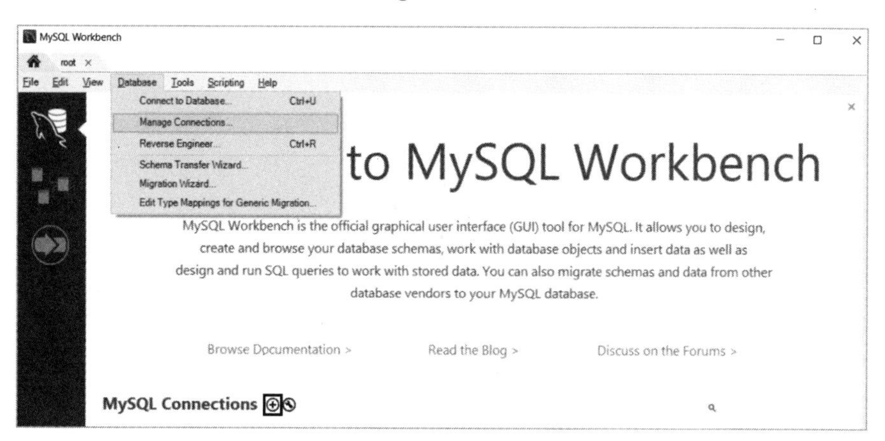

Figure 18-1: Creating a MySQL Workbench connection

The Setup New Connection window will open, as shown in Figure 18-2. Here, enter a connection name (I chose to give the connection the same name as the user: `root`) and enter **`root`** as the username.

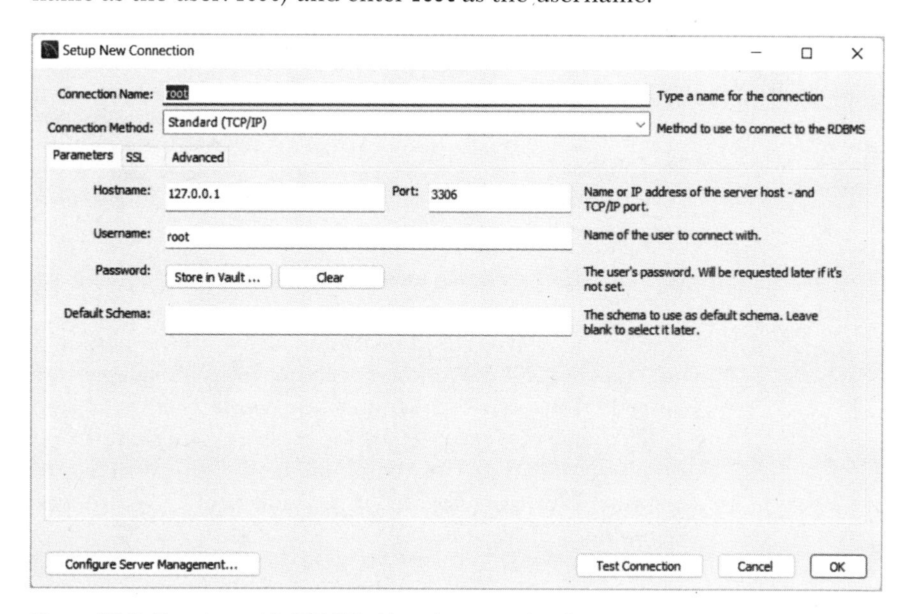

Figure 18-2: Creating a MySQL Workbench connection for `root`

To save the connection, click **OK**. Now you can log in as root in the future simply by clicking the connection.

Since root is a superuser account that has all privileges and can grant privileges to other users, you'll use this connection to run the script to create the database, tables, view, and users for your departments. Figure 18-3 shows the end of that script, but you'll need to run the full one at *https://github.com/ricksilva/mysql_cc/blob/main/chapter_18.sql.*

```
Query 1
Limit to 1000 rows
21 • insert into employee(first_name, last_name, department, job_title, salary)
22   values ('Ann', 'Ellis', 'Technology', 'Programmer', 119500);
23
24 • insert into employee(first_name, last_name, department, job_title, salary)
25   values ('Todd', 'Lynch', 'Legal', 'Compliance Manager', 157000);
26
27 • create view v_employee as
28   select  employee_id,
29           first_name,
30           last_name,
31           department,
32           job_title
33   from    employee;
34
35 • create user accounting_user@localhost identified by 'accounting_password';
36 • create user marketing_user@localhost identified by 'marketing_password';
37 • create user technology_user@localhost identified by 'technology_password';
38 • create user legal_user@localhost identified by 'legal_password';
39 • create user hr_user@localhost identified by 'hr_password';
40
41 • grant select, delete, insert, update on business.employee to hr_user@localhost;
42
43 • grant select on business.v_employee to hr_user@localhost;
44 • grant select on business.v_employee to accounting_user@localhost;
45 • grant select on business.v_employee to marketing_user@localhost;
46 • grant select on business.v_employee to legal_user@localhost;
47 • grant select on business.v_employee to technology_user@localhost;
48
```

Figure 18-3: Creating tables, view, and users and granting access using MySQL Workbench

Now that you've run the script to create usernames for your departments, you'll create MySQL Workbench connections for hr_user and accounting_user. Figure 18-4 shows how to set up a new connection for hr_user.

To create the connection for hr_user, you entered a connection name and username of hr_user. You'll create a connection for accounting_user the same way, using accounting_user for both the connection name and username.

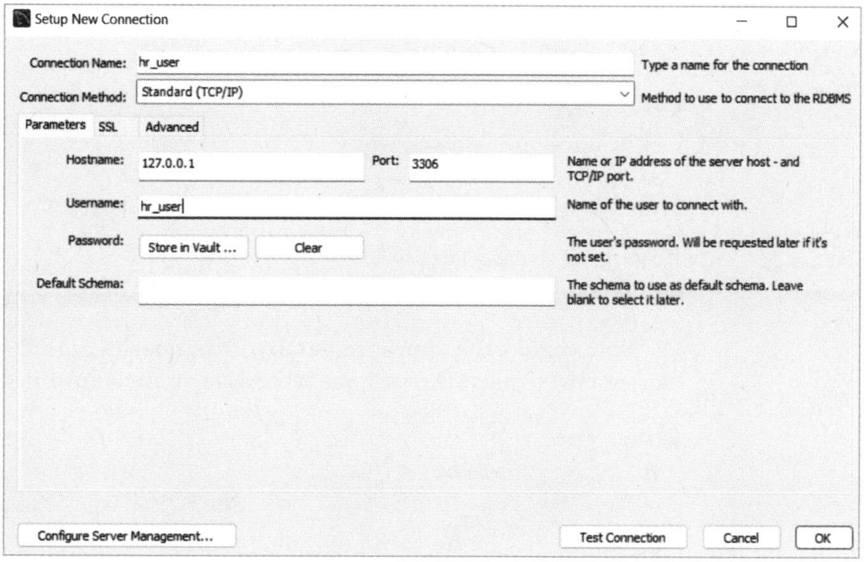

Figure 18-4: Creating a MySQL Workbench connection for `hr_user`

Now you have three connections in MySQL Workbench that you can use, as shown in Figure 18-5.

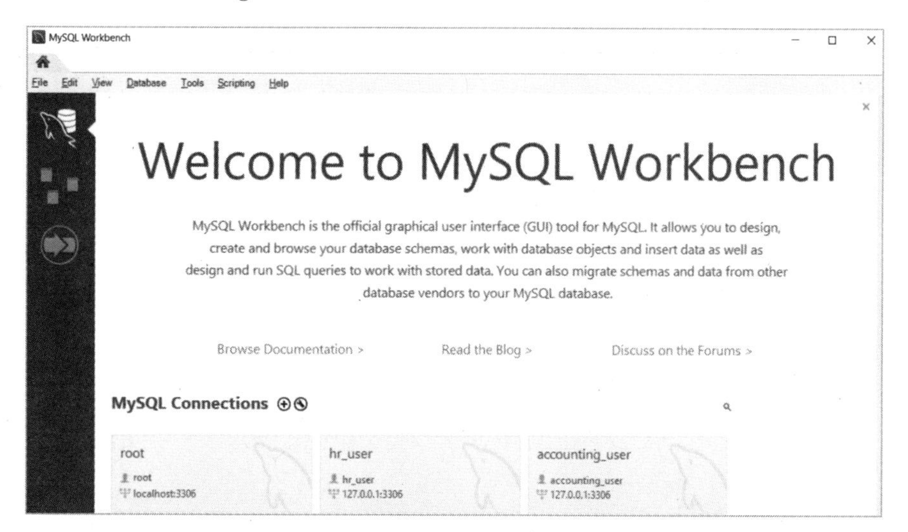

Figure 18-5: MySQL Workbench connections for `root`*,* `hr_user`*, and* `accounting_user`

The connections appear with the names you used when you created them. You can log in to MySQL as each user by clicking the corresponding connection.

You can also open multiple connections at once. Open a connection as `hr_user`, then click the home icon at the top left to return to the welcome screen. From here, open another connection as `accounting_user` by clicking its connection.

You now should see two tabs in MySQL Workbench, labeled `hr_user` and `accounting_user`, as shown in Figure 18-6.

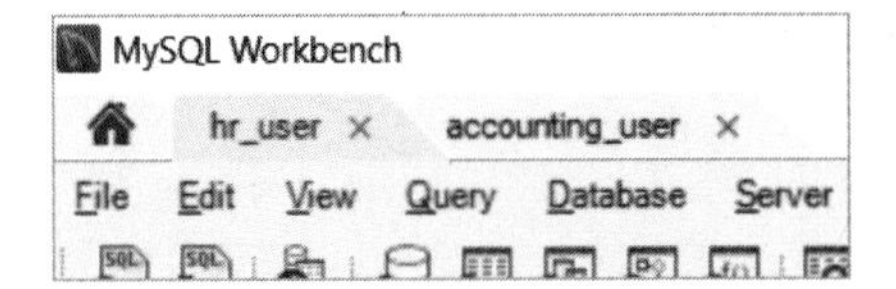

Figure 18-6: You can have multiple connections open in MySQL Workbench.

Simply click the appropriate tab to run queries as that user. Click the `hr_user` tab to query the `employee` table as `hr_user` (Figure 18-7).

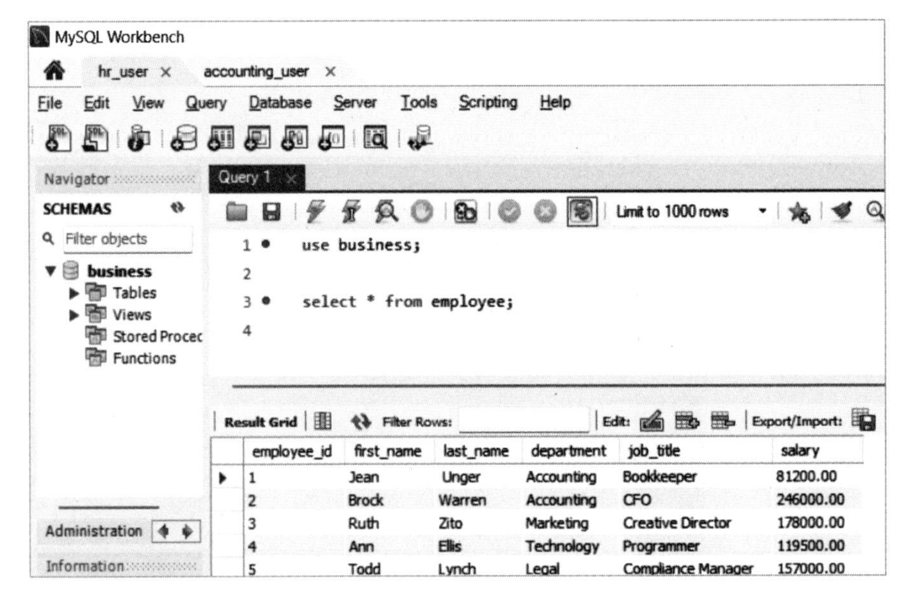

Figure 18-7: Querying the `employee` table as `hr_user`

Now, click the `accounting_user` tab and query the `employee` table again, as shown in Figure 18-8.

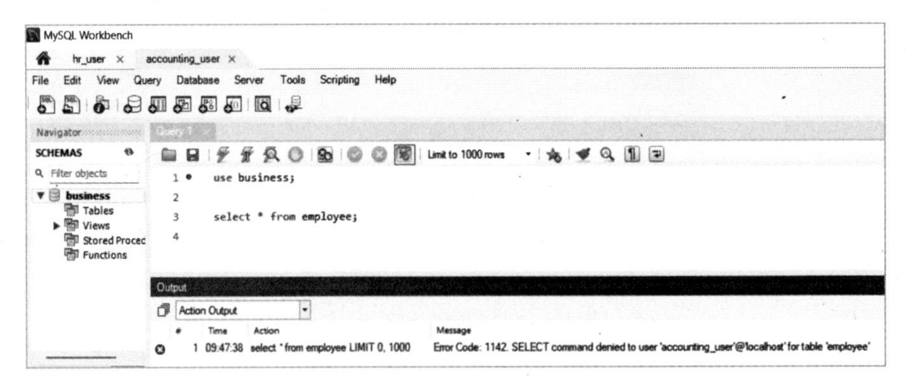

Figure 18-8: The `accounting_user` cannot view the `employee` table.

Because you as root haven't granted access on the employee table to accounting_user, the error SELECT command denied is returned. The accounting_user can, however, select from the v_employee view, so the user can see employee data without the salaries (Figure 18-9).

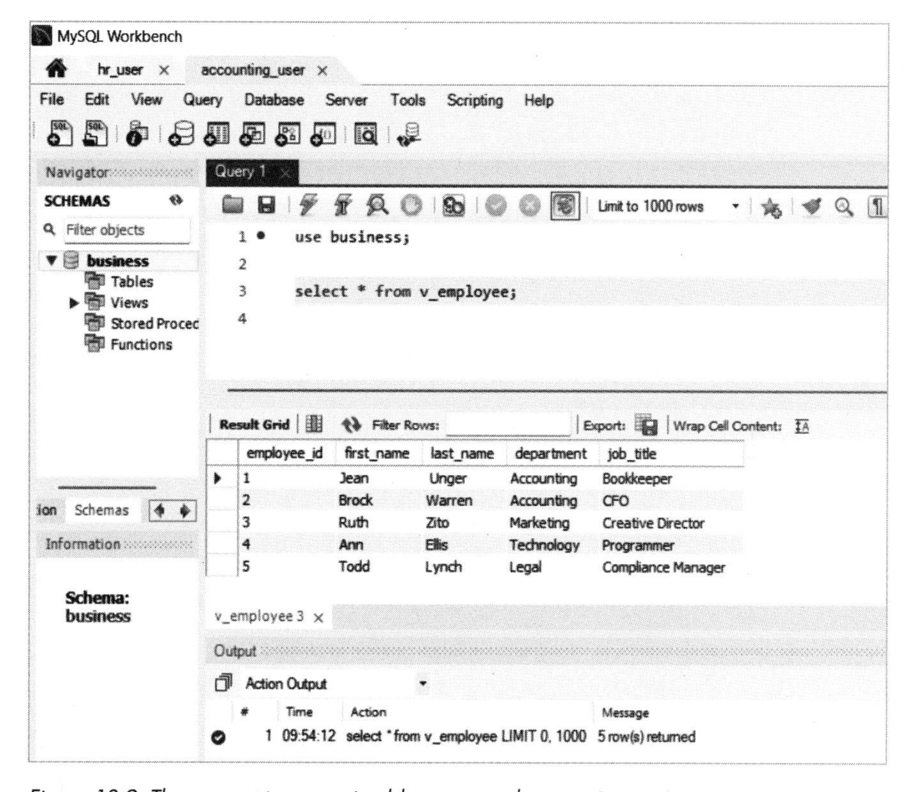

Figure 18-9: The accounting_user is able to query the v_employee view.

Your other database users have the same privileges as accounting_user, meaning they can't query the employee table either, because you haven't granted them access.

TRY IT YOURSELF

18-1. Log in as root and create a view called v_employee_fn_dept that has just the first_name and department columns from the employee table. Query the view. Do you see the results you expected?

An Alternative Approach

There's another way to hide data from particular users. MySQL allows you to grant permissions at the column level; for example, you could grant the select privilege on all the columns in the employee table except for salary:

```
grant select(employee_id, first_name, last_name, department, job_title)
on employee
to technology_user@localhost;
```

This allows technology_user to select any or all of the employee_id, first_name, last_name, department, or job_title columns from the table, like so:

```
select employee_id,
       first_name,
       last_name
from   employee;
```

The result is:

```
employee_id  first_name  last_name
-----------  ----------  ---------
     1       Jean        Unger
     2       Brock       Warren
     3       Ruth        Zito
     4       Ann         Ellis
     5       Todd        Lynch
```

Since you haven't granted select access on the salary column, MySQL will prevent technology_user from selecting that column:

```
select salary
from   employee;
```

The result is an error message:

```
SELECT command denied to user 'technology_user'@'localhost' for table 'employee'
```

If technology_user tries to select all columns using the * wildcard, they will receive the same error message, because they cannot return the salary column. For this reason, I don't favor this approach, as it can lead to confusion. It's more straightforward to allow users to access all permissible tables through a view.

Summary

In this project, you used a view to hide salary information from particular users. This technique could be used to hide any kind of sensitive data in your tables. You also learned how granting and revoking privileges for database users can help to create secure databases by exposing certain pieces of data to specific users.

With these three projects under your belt, you'll be able to build your own databases, load data from files, create triggers to maintain the quality of your data, and use views to protect sensitive data.

Good luck on the next stage of your MySQL journey!

AFTERWORD

Congratulations! You've learned a great deal about MySQL development and you've applied that new knowledge to real-world projects.

One concept I hope has come through in this book is that there are many ways to skin a MySQL cat. There's plenty of room for creativity in designing and developing databases. You can use your knowledge of MySQL to build highly customized systems for your particular needs and interests, from your favorite baseball statistics to your startup's customer list to a massive web-facing database of corporate takeovers.

To dig deeper into MySQL development, you can load public datasets of interest into your own MySQL database. Websites like *https://data.gov* and *https://www.kaggle.com* contain data that's free to use. (Remember to check the terms of the particular datasets you'd like to work with, however.) See if you can load datasets from different sources into MySQL tables and join them in a way that produces new or interesting insights.

I congratulate you on how far you've come. Learning to think in rows and columns is no small feat. I hope you'll continue to take the time to learn new skills all throughout your life. Knowledge is definitely power.

INDEX

Symbols

A

B

E

L

M

N

O

P

T

U

V

W

Y

MySQL Crash Course is set in New Baskerville, Futura, Dogma, and TheSansMono Condensed.

RESOURCES

Visit *https://nostarch.com/mysql-crash-course* for errata and more information.